STUDENT UNIT GUIDE

NEW EDITION

Edexcel A2 Biology Unit 4

The Natural Environment and Species Survival

Mary Jones

PHILIP ALLAN

Philip Allan, an imprint of Hodder Education, an Hachette UK company, Market Place, Deddington, Oxfordshire OX15 0SE

Orders
Bookpoint Ltd, 130 Milton Park, Abingdon, Oxfordshire OX14 4SB
tel: 01235 827827
fax: 01235 400401
e-mail: education@bookpoint.co.uk
Lines are open 9.00 a.m.–5.00 p.m., Monday to Saturday, with a 24-hour message answering service. You can also order through the Philip Allan website: www.philipallan.co.uk

ISBN 978-1-4441-7297-3

First printed 2012
Impression number 5 4 3 2
Year 2017 2016 2015 2014

Cover photo: Fotolia

Printed in Dubai

Typeset by Greenhill Wood Studios

Hachette UK's policy is to use papers that are natural, renewable and recyclable products and made from wood grown in sustainable forests. The logging and manufacturing processes are expected to conform to the environmental regulations of the country of origin.

Contents

Content Guidance

Questions & Answers

Getting the most from this book

Examiner tips
Advice from the examiner on key points in the text to help you learn and recall unit content, avoid pitfalls, and polish your exam technique in order to boost your grade.

Knowledge check
Rapid-fire questions throughout the Content Guidance section to check your understanding.

Knowledge check answers
I Turn to the back of the book for the Knowledge check answers.

Summary

Summaries
● Each core topic is rounded off by a bullet-list summary for quick-check reference of what you need to know.

Questions & Answers

Exam-style questions

Examiner comments on the questions
Tips on what you need to do to gain full marks, indicated by the icon **e**.

Sample student answers
Practise the questions, then look at the student answers that follow each set of questions.

Examiner commentary on sample student answers
Find out how many marks each answer would be awarded in the exam and then read the examiner comments (preceded by the icon **e**) following each student answer.

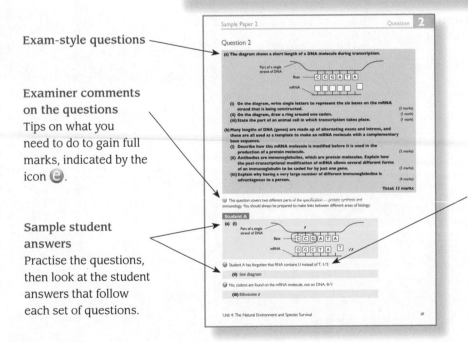

About this book

This book is the third in a series of five, which will help you to prepare for the Edexcel A-level biology examination. It covers **Unit 4: The Natural Environment and Species Survival**. This is the first of two content-based units that make up the A2 biology examination. The other four books in the series cover Units 1, 2, 5 and 3/6.

This guide has two main sections:
- **Content Guidance** This provides a summary of the facts and concepts that you need to know for the Unit 4 examination.
- **Questions and Answers** This section contains two specimen papers for you to try, each worth 90 marks. There are also two sets of answers for each question, one from a student who is likely to get a C grade and another from a student who is likely to get an A grade.

The specification

It is a good idea to have your own copy of the Edexcel biology specification. It is you who is going to take this examination, not your teacher, and so it is your responsibility to make sure you know as much about the exam as possible. You can download a copy free from **www.edexcel.org.uk**.

The A2 examination is made up of three units:

- Unit 4   The Natural Environment and Species Survival
- Unit 5   Energy, Exercise and Coordination
- Unit 6   Practical Biology and Investigative Skills

This book covers Unit 4. There is also a practical guide that covers AS Unit 3 and A2 Unit 6 and is based on practical work that you will do in your biology classes.

Unit 4 content

The content of each unit is clearly set out in the specification. Unit 4 has two topics:

- On the wild side
- Infection, immunity and forensics

On the wild side looks at the ways in which organisms obtain energy from their environment, and use it to make ATP in their cells. Plants transfer energy from sunlight into organic compounds by photosynthesis, and the rate at which they do this is known as productivity. Some of this energy can be transferred along food chains to consumers. Much energy is lost in this process, and you need to be able to calculate the efficiency of energy transfer. We look at how environmental factors affect the distribution and abundance of species in a habitat, and how this can change over time.

This leads us into a brief review of the carbon cycle, concentrating on the roles of carbon dioxide and methane in the greenhouse effect, and how increases in the concentrations of these gases in the atmosphere are contributing to global warming. You need to be aware of some of the evidence for these processes, and of how data can be used to construct models that help us to predict what may happen to global climate in the future.

Organisms are going to have to adapt to these changes in their environment, and natural selection will play an important role in this. We look at how evolution can lead to speciation, and the role of the scientific community in evaluating new evidence that may throw light on this process.

Infection, immunity and forensics begins with a very brief review of DNA structure, and then looks in detail at how the genetic code is used to synthesise proteins in cells. A simple outline of DNA profiling comes next, including some practical details of the polymerase chain reaction (for amplifying small samples of DNA) and the use of electrophoresis in separating DNA fragments of different lengths.

The structure of bacteria and viruses is revisited, followed by consideration of how pathogenic micro-organisms get into the human body, and the way we respond to such infections. We consider the evolutionary race between pathogens and their hosts, and how antibiotics can be used to cure bacterial infections.

Finally, we look briefly at how succession on a corpse, as well as other changes taking place in it, can be used to estimate the time of death.

Unit 4 assessment

Unit 4 is assessed in an examination lasting 1 hour 30 minutes. The questions are all structured — that is, they are broken up into several parts, with spaces in which you write your answers. There are 90 marks available on the paper.

What is assessed?

It's easy to forget that your examination isn't just testing what you *know* about biology — it's also testing your *skills*. It's difficult to overemphasise how important these are.

The Edexcel examination tests three different assessment objectives (AOs). The following table gives a breakdown of the proportion of marks awarded to knowledge and to skills in the A2 examination:

Assessment objective	Outline of what is tested	Percentage of marks
AO1	Knowledge and understanding of science and of How Science Works	26–30
AO2	Application of knowledge and understanding of science and of How Science Works	42–48
AO3	How Science Works	26

AO1 is about remembering and understanding all the biological facts and concepts you have covered in this unit. AO2 is about being able to *use* these facts and concepts in new situations. The examination paper will include questions that contain unfamiliar contexts or sets of data, which you will need to interpret in the light of the biological knowledge you have. When you are revising, it is important that you try to develop your ability to do this, as well as just learning the facts.

AO3 is about How Science Works. Note that this comes into AO1 and AO2 as well. A science subject such as biology is not just a body of knowledge. Scientists do research to find out how things around them work, and new research continues to find out new things all the time. Sometimes new research means that we have to change our ideas. For example, not all that long ago people were encouraged to eat lots of eggs and drink lots of milk, because it was thought to be 'healthy'. Now we know we need to take care not to eat too many animal-based fats, because new research has found links between a fatty diet and heart disease.

How Science Works is about developing theories and models in biology, and testing them. It involves doing experiments to test hypotheses, and analysing the results to determine whether the hypotheses are supported or disproved. You need to appreciate why science does not always give us clear answers to the questions we ask, and how we can design good experiments whose results we can trust.

Scientific language

Throughout your biology course, and especially in your examination, it is important to use clear and correct biological language. Scientists take great care to use language precisely. If doctors or researchers do not use exactly the correct words when communicating with someone, then what they are saying could easily be misinterpreted. Biology has a huge number of specialist terms (probably more than any other subject you can choose to study at AS) and it is important that you learn them and use them. Your everyday conversational language, or what you read in the newspaper or hear on the radio, is often not the kind of language required in a biology examination. Be precise and careful in what you write, so that an examiner cannot possibly misunderstand you.

The examination

Time

You will have 90 minutes to answer questions worth 90 marks. That gives you 1 minute per mark. When you are trying out a test question, time yourself. Are you working too fast? Or are you taking too long? Get used to what it feels like to work at just over a-mark-a-minute rate.

It's not a bad idea to spend one of those minutes just skimming through the exam paper before you start writing. Maybe one of the questions looks as though it is going to need a bit more of your time than the others. If so, make sure you leave a little bit of extra time for it.

Read the question carefully

That sounds obvious, but students lose large numbers of marks by not doing it.

- There is often vital information at the start of the question that you'll need in order to answer the questions themselves. Don't just jump straight to the first place where there are answer lines and start writing. Start reading at the beginning! Examiners are usually careful not to give you unnecessary information, so if it

is there it is probably needed. You may like to use a highlighter to pick out any particularly important bits of information in the question.

- Do look carefully at the command words (the ones right at the start of the question) and do what they say. For example, if you are asked to explain something then you won't get many marks — perhaps none at all — if you describe it instead. You can find all these words in an appendix near the end of the specification document.

Depth and length of answer

The examiners will give you two useful guidelines about how much you need to write.

- **The number of marks**. Obviously, the more marks the more information you need to give. If there are 2 marks, then you'll need to give two different pieces of information in order to get both of them. If there are 5 marks, you'll need to write much more.
- **The number of lines**. This isn't such a useful guideline as the number of marks, but it can still help you to know how much to write. If you find your answer won't fit on the lines, then you probably haven't focused sharply enough on the question. The best answers are short and precise.

Writing, spelling and grammar

The examiners are testing your biology knowledge and skills, not your English skills. Still, if they can't understand what you have written then they can't give you any marks. It's your responsibility to communicate clearly — don't scribble so fast that the examiner cannot read what you have written.

In general, incorrect spellings are not penalised. If the examiner knows what you are trying to say then he or she will give you credit. However, if your wrongly spelt word could be confused with another, then you won't be given the mark. For example, if you write 'meitosis', then the examiner can't know whether you mean meiosis or mitosis, so you'll be marked wrong.

Like spelling, bad grammar isn't taken into account. Once again, though, if it is so bad that the examiner cannot understand you, then you won't get marks. A common problem is to use the word 'it' in such as way that the examiner can't be certain what 'it' refers to. A good general rule is never to use this word in an exam answer.

Content Guidance

On the wild side

Energy and living organisms

ATP

ATP stands for adenosine triphosphate. ATP is a phosphorylated nucleotide — it has a similar structure to the nucleotides that make up RNA. However, it has three phosphate groups attached to it instead of one.

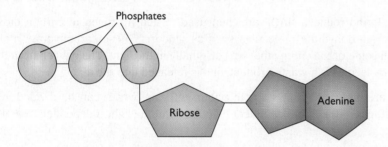

Figure 1 An ATP molecule

ATP is used as the energy currency in every living cell. ATP molecules contain a lot of chemical energy. When an ATP molecule is hydrolysed, losing one of its phosphate groups, some of this energy is released and can be used by the cell. In this process, the ATP is converted to ADP (adenosine diphosphate).

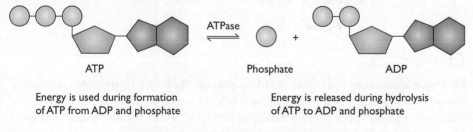

Energy is used during formation of ATP from ADP and phosphate

Energy is released during hydrolysis of ATP to ADP and phosphate

Figure 2 Hydrolysis and formation of ATP

Cells use energy for many different purposes, such as for the synthesis of proteins and other large molecules, for active transport, for the transmission of nerve impulses or for the contraction of muscles. Each cell makes its own ATP. The hydrolysis of one ATP molecule releases a small 'packet' of energy that is often just the right size to fuel a particular step in a process. A glucose molecule, on the other hand, would contain

Knowledge check 1

How is the structure of an ATP molecule similar to an RNA nucleotide? How does it differ?

far too much energy, so a lot would be wasted if cells used glucose molecules as their immediate source of energy.

All cells make ATP by respiration. This is described in Unit 5. Some plant cells also make ATP during some of the reactions of photosynthesis, although most of this ATP is immediately used during other steps in photosynthesis.

Photosynthesis

Photosynthesis is a series of reactions in which light energy is converted to chemical energy. Light energy is trapped by chlorophyll, and this energy is then used to:

- split apart the strong bonds in water molecules to release hydrogen
- produce ATP
- reduce a substance called NADP

NADP stands for nicotinamide adenine dinucleotide phosphate, which is a coenzyme (that is, a molecule required for an enzyme to be able to catalyse a reaction). The term 'reduce' means to add hydrogen, so reduced NADP has had hydrogen added to it.

The ATP and reduced NADP are then used to add hydrogen to carbon dioxide to produce carbohydrate molecules such as glucose. These carbohydrate molecules contain some of the energy that was originally in the light. The oxygen from the split water molecules is a waste product, and is released into the air.

There are many different steps in photosynthesis, which can be divided into two main stages — the light-dependent reactions and the light-independent reactions.

Knowledge check 2

Students sometimes write 'in photosynthesis, carbon dioxide is changed into oxygen'. Why is this wrong?

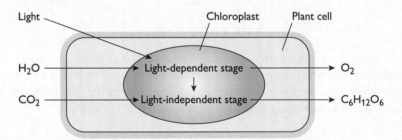

Figure 3 An overview of photosynthesis

Chloroplasts

Photosynthesis takes place inside **chloroplasts**. These are organelles surrounded by two membranes, called an **envelope**. They are found in mesophyll cells in leaves. Palisade mesophyll cells contain most chloroplasts but they are also found in spongy mesophyll cells. Guard cells also contain chloroplasts.

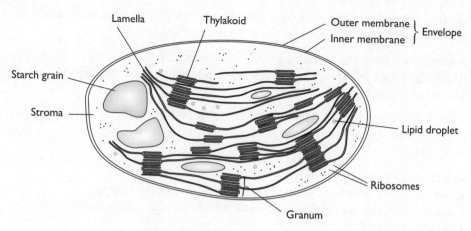

Figure 4 Structure of a chloroplast

The membranes inside a chloroplast are called **lamellae**, and it is here that the light-dependent reactions take place. The membranes contain **chlorophyll** molecules, arranged in groups called **photosystems**. There are two kinds of photosystems, PSI and PSII, each of which contains slightly different kinds of chlorophyll.

There are enclosed spaces between pairs of membranes, forming fluid-filled sacs called **thylakoids**. These are involved in photophosphorylation — the formation of ATP using energy from light. Thylakoids are often arranged in stacks called **grana** (singular: granum).

The 'background material' of the chloroplast is called the **stroma**, and this is where the light-independent reactions take place.

Chloroplasts often contain starch grains and lipid droplets. These are stores of energy-containing substances that have been made in the chloroplast but are not immediately needed by the cell or by other parts of the plant.

The light-dependent reactions

Chlorophyll molecules in PSI and PSII absorb light energy. The energy excites electrons, raising their energy level so that they leave the chlorophyll.

PSII contains an enzyme that splits water when activated by light. This reaction is called photolysis ('splitting by light'). The water molecules are split into oxygen and hydrogen atoms. Each hydrogen atom then loses its electron, to become a positively charged hydrogen ion (proton), H^+. The electrons are picked up by the chlorophyll in PSII, to replace the electrons it lost. The oxygen atoms join together to form oxygen molecules, which diffuse out of the chloroplast and into the air around the leaf.

$$2H_2O \xrightarrow{\text{light}} 4H^+ + 4e^- + O_2$$

The electrons emitted from PSII are picked up by electron carriers in the membranes of the thylakoids. They are passed along a chain of these carriers, losing energy as they go. The energy they lose is used to make ADP combine with a phosphate group, producing ATP. This is called **photophosphorylation**. At the end of the

Examiner tip

The supply of energy for the light-dependent reactions comes from light. These reactions, therefore, are not affected by temperature, unlike most other reactions in living organisms.

electron carrier chain, the electron is picked up by PSI. This replaces the electron the chlorophyll in PSI had lost when it absorbed the light energy.

The electrons from PSI are passed along a different chain of carriers to NADP. The NADP also picks up the hydrogen ions from the split water molecules. The NADP becomes reduced NADP.

We can show all of this in a diagram called the **Z-scheme**. The higher up the diagram, the higher the energy level. If you follow one electron from a water molecule, you can see how it:

- is taken up by PSII
- has its energy raised as the chlorophyll in PSII absorbs light energy
- loses some of this energy as it passes along the electron carrier chain
- is taken up by PSI
- has its energy raised again as the chlorophyll in PSI absorbs light energy
- becomes part of a reduced NADP molecule

Knowledge check 3

The Z-scheme shows that electrons lose energy as they pass along the chains of electron carriers. Where does this energy go?

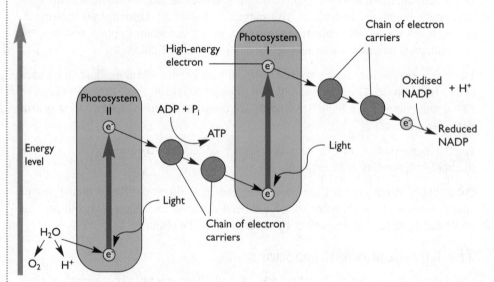

Figure 5 Summary of the light-dependent reactions of photosynthesis — the Z-scheme

At the end of this process, two new high-energy substances have been made. These are ATP and reduced NADP. Both of them will now be used in the next stage of photosynthesis, the light-independent reactions.

The light-independent reactions

These take place in the stroma of the chloroplast, where the enzyme ribulose bisphosphate carboxylase, usually known as **RUBISCO**, is found.

Knowledge check 4

What are the substrates and product of the reaction catalysed by RUBISCO?

Carbon dioxide diffuses into the stroma from the air spaces within the leaf. It enters the active site of RUBISCO, which combines it with a 5-carbon compound called ribulose bisphosphate, **RuBP**. The products of this reaction are two 3-carbon molecules, glycerate 3-phosphate, **GP**. The combination of carbon dioxide with RuBP is called **carbon fixation**.

Energy from ATP and hydrogen from reduced NADP are then used to convert the GP into triose phosphate, **TP**. (This is sometimes known as glyceraldehyde 3-phosphate, **GALP** for short.) This is the first carbohydrate produced in photosynthesis.

Most of the triose phosphate is used to produce ribulose bisphosphate, so that more carbon dioxide can be fixed. The rest is used to make glucose or whatever other organic substances the plant cell requires. These include polysaccharides such as starch for energy storage and cellulose for making cell walls, sucrose for transport, amino acids for making proteins, lipids for energy storage and nucleotides for making DNA and RNA.

This cyclical series of reactions is known as the **Calvin cycle**.

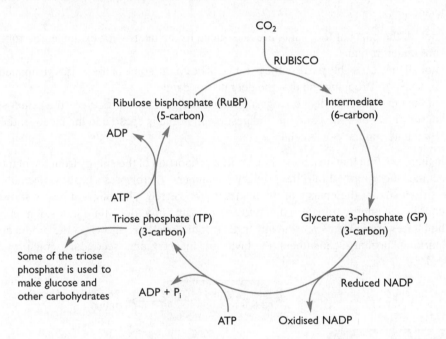

Figure 6 The Calvin cycle

Examiner tip

Some old textbooks refer to the light-independent reactions as the 'dark reactions'. Do not use this term. The reactions can happen perfectly well in light — they just don't need light.

Knowledge check 5

Where do the reduced NADP and the ATP, used in the Calvin cycle, come from?

Productivity

The rate at which a plant fixes carbon, producing organic compounds such as carbohydrates, is known as its **productivity**. Productivity is generally measured as the amount of energy in kilojoules in the organic compounds fixed by a certain area of plants (for example $1\,m^2$) per year. The units are therefore $kJ\,m^{-2}\,y^{-1}$.

The total amount of light energy transferred into the plant by photosynthesis is gross primary productivity, **GPP**.

The plant uses some of this energy itself. It breaks down carbohydrates (and other organic compounds) by respiration, using the energy to fuel its own body processes such as active transport. This energy is eventually lost from the plant as heat. The total amount of energy left in the plant tissues is therefore the GPP minus this heat energy lost through respiration. The energy left in the plant is known as net primary productivity, **NPP**.

NPP = GPP – respiration

Energy transfer in food chains

Animals and many other organisms (fungi and other decomposers) depend on the energy fixed by plants to provide them with *their* energy. The energy theoretically available to them is the NPP of the plants in the ecosystem.

However, imagine a population of deer feeding on an area of grassland. Not all the energy in the plant tissues actually enters the deer population. Reasons for this include:

- not all the parts of the plants are accessible to the deer — for example, the roots are underground
- not all the accessible parts of the plants will be eaten; some of the grass is trampled, or soiled by droppings so that the deer do not eat it
- of the grass that the deer do ingest, much is indigestible (particularly the cellulose in the plant cell walls) and this indigestible material is egested in the faeces rather than entering the deer's cells

We therefore find that, in a food chain, only a proportion of the energy from the plants becomes incorporated into the primary consumers' (herbivores') tissues. Normally, less than 10% of the energy in the plants is transferred. The diagram below shows quantities of energy transferred between organisms in a food chain in a salt marsh. The figures are in $kJ\,m^{-2}\,year^{-1}$. Only three trophic levels are shown — the producers (*Spartina*), primary consumers (herbivorous insects) and secondary consumers (spiders).

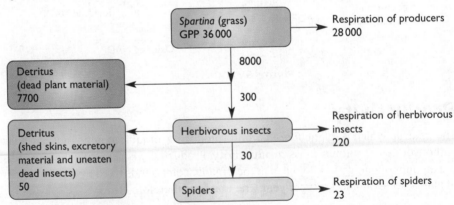

Figure 7 Energy transfer in a salt marsh food chain

We can use this diagram to calculate the NPP of the producers and also the efficiency of energy transfer between the trophic levels.

NPP = GPP – R
= 36 000 – 28 000 = 8000

This is the quantity of energy actually stored in the tissues of the producers. You can see from the diagram that this is the quantity of energy that is available to be passed on from the plants to the next trophic level.

However, you can also see that only $300 \, kJ \, m^{-2} \, year^{-1}$ is actually passed on to the next trophic level.

$$\frac{\text{efficiency of}}{\text{energy transfer}} = \frac{\text{energy in one trophic level}}{\text{energy in previous trophic level}} \times 100$$

so efficiency of energy transfer from producers to primary consumers

$$= \frac{300}{8000} \times 100 = 3.75\%$$

Distribution and abundance of organisms

The places in which a particular species can be found, and the numbers of that species, are affected by aspects of the environment. These are called environmental factors. They can be classified into **biotic** factors (caused by other organisms) and **abiotic** factors (caused by non-living features of the environment).

Biotic factors include predation, competition (both between and within a species) and availability of food (for consumers). Abiotic factors include light intensity, temperature, water supply, wind speed, and factors relating to the soil, known as edaphic factors. These include mineral composition of the soil, pH, drainage and water-holding capacity, and the quantity of humus present.

Investigating distribution and abundance of organisms in a habitat

You can choose almost any habitat for your investigation. You may need to study two different habitats in order to use all of the techniques and factors listed in the specification. It is often most interesting to investigate how the distribution and abundance of organisms change in an area where there are spatial differences in environmental factors, such as on a sea shore. You could also compare two similar habitats in which at least one factor differs. The description below suggests how you could investigate the distribution of a species such as dandelions in open ground and woodland.

First, choose a place where there is a grassy area open to full sunlight, which grades into woodland where the ground is heavily shaded by trees. Look around the habitat and get the 'feel' of what is growing where and any obvious differences in environmental factors. If possible, construct a hypothesis that you will test, for example: The distribution and abundance of dandelions in this area is related to the light intensity near the soil surface; where light intensity is low, there are few or no dandelions.

Use pegs and string to mark out a line, starting in the open area and ending in the shady area, along which you will make your measurements and collect your data.

Knowledge check 7

Calculate the efficiency of energy transfer between primary consumers and secondary consumers.

This is called a **transect**. Mark off equal distance intervals along the line — for example, every 1 m.

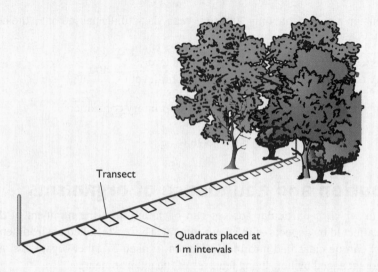

Figure 8 A typical line transect

Measure abiotic factors at each 1 m point along the line. These could include light intensity (using a light meter), air temperature and soil temperature (thermometer), wind speed (anemometer), humidity (hygrometer) and soil pH (pH meter). If time permits, repeat these measurements each day over a period of at least 1 week, and also at different times of day.

Sample the organisms along the transect. Place a 1 m² quadrat at one end of the transect. Count the number of dandelion plants inside the quadrat and record them. Repeat at each 1 m interval along the quadrat. Alternatively, depending on the hypothesis you are testing, you could identify all the different species of plants inside the quadrat and estimate the percentage of the area of the quadrat covered by each species. (The percentage could add up to more than 100%, because some plants may overlie others.)

Display your data so that you can see any correlations between the numbers of dandelions at different points along the transect (their abundance) and any differences in the abiotic factors that you have measured.

To ensure that your data are as reliable as possible:
- record abundance in at least 15 quadrats along the transect and preferably more
- carry out at least one repeat investigation with another transect placed parallel to the first one; you can then calculate mean values for plant abundance in the quadrats at each distance along the transect

The niche concept

Knowledge check 8

Define the terms 'habitat' and 'community'.

In your AS course, you saw that each species of organism has its own particular role to play within a community of living organisms. This is called its **niche**. For example, dandelions require light in order to be able to photosynthesise, and are adapted for growing where light is fairly bright. They require moist soil, but cannot grow where the soil is waterlogged. They do not thrive in shaded woodland conditions. Dandelion leaves and flowers are eaten by many different herbivores, and they are able to grow very close to the ground so that few herbivores are able to eat absolutely all of their leaves. Their fruits (dandelion clocks) contain seeds that are dispersed by wind.

There is no other species of plant whose niche exactly matches this niche. Niches of different species living in the same habitat frequently overlap one another, but they are always found to have significant differences between them.

The niche of an organism partly determines its distribution and abundance within a particular habitat. For example, dandelions are not found in deep woodland because they require plenty of light. They can grow on closely-mown lawns because they are able to grow in very low rosettes, which avoid having their growing points cut off by the mower.

Succession

If the abiotic or biotic factors in a habitat change over time, this will result in changes in the distribution and abundance of species in the community. A gradual change in ecological factors and the community over time is known as **succession**.

For example, as global temperatures increase, many glaciers are retreating. A glacier is a slow-moving river of ice. Glaciers retreat as the lowest parts of them (where they enter the sea or a lake) melt.

The moving ice scoured the ground beneath it, destroying all life. So when the ice melts, it leaves bare ground. Over time, different species of organisms colonise this ground.

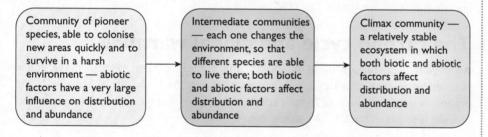

Figure 9 Succession to a climax community

The first species to live there are called **pioneer species**. These are organisms which are able to disperse quickly to new areas, and that can live in the harsh conditions. Pioneer species include many lichens, and also some species of flowering plants such as willowherb and mountain avens. These have fluffy seeds that are blown over quite large distances by the wind, and they are able to grow in the thin, poor soil that has been left exposed by the melting ice. The newly-exposed soil is very poor in nitrate or ammonium ions, and several pioneer species are nitrogen-fixing flowering plants, such as mountain avens. These have *Rhizobium* or other nitrogen-fixing bacteria in their roots, which are able to use nitrogen from the air to make ammonium ions. The plant can then use these to make amino acids and proteins. Still, conditions

are tough, as there is no shelter from drying winds, and the soil contains little or no humus so tends to dry out or become waterlogged very easily.

These pioneer species alter the environmental conditions in the habitat. When they die, their remains form humus which helps to improve the soil. They provide shelter in which seeds of other plants can germinate and get a foothold for their roots. They add ammonium or nitrate ions to the soil. They provide food for animals. Gradually, over time, more and more species arrive in the habitat and are able to live there.

About 50 years after the glacier retreated, various shrubs and small trees, such as alders, are found growing in the habitat. Eventually, the community settles down to a forest of tall hemlock and sitka spruce trees, in which many other plants and a wide range of animal species live. This community does not change, and it is called the **climax community**.

Examiner tip

When you are studying succession by recording distribution of organisms in *space*, it is important to remember that succession is about changes over *time*.

We can study succession by recording the species found in a particular area at regular intervals over a long period of time. However, it is often possible to see many different stages of the succession at one moment in time. For example, if a glacier is steadily retreating, then we can find the earliest stages of the succession just at the edge of the glacier, and the later stages further away from it, because these were first exposed many years ago. So we can study changes in a community in *space* (perhaps using a transect) to tell us how it has been changing with *time*.

In the early stages of a succession, the distribution and abundance of organisms is largely determined by abiotic factors, such as soil conditions (or the lack of soil), availability of water, light, temperature and so on. As the succession progresses, biotic factors become more important. For example, the early pioneer species, which are well adapted for growing in harsh conditions, disappear as time goes on, because they are not able to compete with other species of plants that appear later in the succession.

The carbon cycle and global warming

All organic substances in living organisms contain carbon. Plants make these substances in photosynthesis, using carbon dioxide from the air. When any living organism respires, some of these carbon compounds are broken down and carbon dioxide is released back to the atmosphere.

Decomposers use dead organisms and their waste materials as energy sources. These decomposers respire, again releasing carbon dioxide to the air.

In the Carboniferous period, much of the land on Earth was waterlogged. When plants died, the lack of oxygen in the soil meant that decomposers could not break them down completely. The partly decomposed remains of plants and other organisms became compressed and heated deep in the ground, eventually forming fossil fuels such as coal, oil and natural gas. These contain a great deal of carbon that was originally taken from the atmosphere by photosynthesising plants. When fossil fuels are burnt, the carbon in them is released back to the air as carbon dioxide.

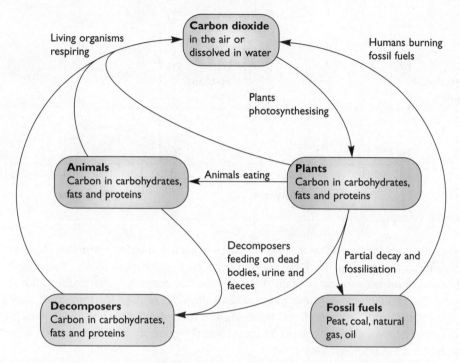

Figure 10 The carbon cycle

Knowledge check 9

Why is photosynthesis such a crucial part of the carbon cycle?

Global warming and its causes

We are currently using up these fossil fuels at a great rate, and this is contributing to a steady rise in the carbon dioxide concentration in the air. The concentration of carbon dioxide in the atmosphere is currently rising at about 1.9 parts per million per year.

Carbon dioxide is a greenhouse gas, trapping long-wavelength radiation and preventing it from escaping from the Earth into space. Another gas, methane, CH_4, is even more efficient at trapping this radiation. The concentrations of both carbon dioxide and methane in the atmosphere are rising.

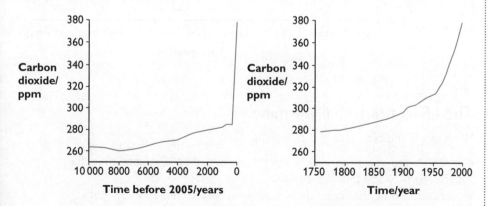

Figure 11 Changes in carbon dioxide concentration in the atmosphere

Examiner tip

Note the different units for carbon dioxide and methane. Although concentrations of methane are much less than carbon dioxide, it is still very significant because it is a more effective greenhouse gas.

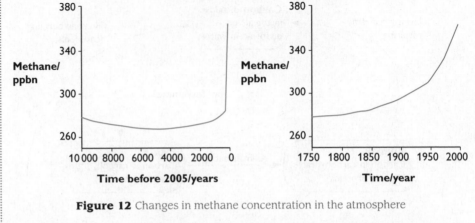

Figure 12 Changes in methane concentration in the atmosphere

Knowledge of the carbon cycle shows us that we could reduce the emissions of carbon dioxide to the atmosphere if we did not burn such large quantities of fossil fuels. Burning biomass — that is, plant material that has recently taken carbon dioxide from the atmosphere — is less harmful, as it does not release the long-term stores of carbon dioxide in fossil fuels. Reducing the rate of deforestation, or reforesting deforested areas, could also help, by increasing the quantity of carbon dioxide taken from the air in photosynthesis.

The mean global temperature is also rising, due at least in part to these increases in greenhouse gases.

Knowledge check 10

Suggest why the global surface temperature has been plotted as 'difference from the 1961–1990 mean temperature' rather than simply as 'mean temperature'.

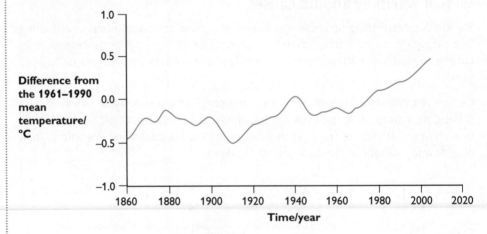

Figure 13 Mean global surface temperature since 1860

The effects of global warming

Mean global temperatures are increasing.

- In 2008, the mean surface temperature was 0.33°C higher than the average between 1850 and the present time.
- The 20th century has been the warmest century in the last 1000 years.
- The period 2001–2008 was 0.19°C warmer than the period 1991–2001.

Edexcel A2 Biology

These rising temperatures are causing many different effects. Sea levels are rising, as ocean water expands with heating, and as glaciers and ice sheets melt. Rainfall patterns are changing, so that some parts of the world are getting less rain and some are getting more. The annual pattern of seasons is changing in many areas, so that spring comes earlier and winter comes later. These changes are affecting plants and animals in many different ways. They include the following:

- Changes in distribution. Plants and animals adapted for living in a particular climate may no longer be able to survive in their current habitats. For example, the Scottish primrose is adapted to a cold climate with short summers, and currently grows only in the very north of Scotland and in the Orkney Isles. As temperatures rise, it may no longer be able to survive there and will have nowhere else to go. Polar bears may no longer have any suitable habitat, as the extent of the Arctic ice sheets decreases; it is predicted that there may eventually be no summer ice at all. Other species, such as Roesel's bush cricket, are adapted to warmer conditions, and these may benefit as they will be able to extend their range in Britain northwards.
- Changes in rate of development and life cycles. For many organisms, warmer temperatures enable them to begin their life cycle earlier in spring, and to complete it more quickly, because their rates of development are greater. For example, aphids (greenfly) are insects that feed by sucking sap from phloem vessels in plants. Many of them are important agricultural pests as they feed on crops such as wheat or fruit trees. They have a life cycle in which flying individuals are produced only in spring. These flying aphids can travel over quite large distances, finding their way to their plant food sources and settling down there to breed. In recent years, flying aphids have been appearing much earlier in the year, almost certainly because temperatures are a little higher. This gives them more time to complete more life cycles during the year, so that population sizes become greater. However, if temperatures continue to rise, the aphids may be unable to survive in their usual habitats because they cannot breed successfully at temperatures much above 28 °C.

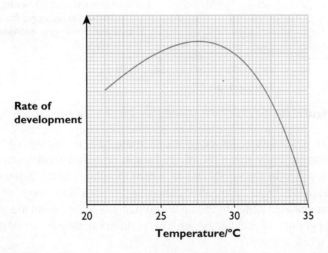

Figure 14 Effect of temperature on the rate of development of a species of aphid

How temperature affects rate of development

Temperature affects the rate of development of living organisms because it affects enzyme activity.

Enzymes are proteins that act as biological catalysts. An enzyme molecule has an active site into which its substrate molecule fits and briefly binds. The substrate molecule is pulled slightly out of shape, causing it to react so that products are formed.

At low temperatures, both the enzyme molecules and the substrate molecules are moving slowly. They bump into each other relatively infrequently and, even when they do collide, they do not have much energy to react.

As temperature increases, the kinetic energy of the molecules increases. They move around faster, and therefore the frequency of collisions between the substrate and enzyme increases. They also collide with more energy. This increases the rate of reaction.

However, above a certain temperature, the hydrogen bonds that help to maintain the tertiary structure (3D shape) of the enzyme molecule begin to break. This causes the active site to lose its shape, so the substrate no longer fits. The higher the temperature, the more hydrogen bonds break and the greater the loss of shape. The enzyme is said to be denatured. It no longer catalyses the reaction.

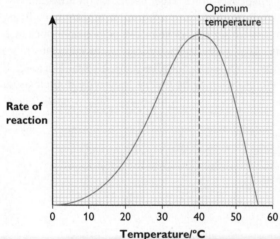

As temperature rises from 0 to about 38 °C, an increase in the kinetic energy of the molecules results in increased collision frequency and therefore increased rate of reaction.

As temperature rises above about 40 °C, hydrogen bonds in the enzyme break so that it loses its shape and the substrate no longer fits in the active site, resulting in decreased rate of reaction.

Figure 15 The effect of temperature on the rate of enzyme activity

The optimum temperature for enzymes in mammals is usually about 38 °C, as this is the normal body temperature. In plants, enzymes tend to have optimum temperatures suited to the conditions in which the plant normally lives, and these may be quite a bit lower than 38 °C. Micro-organisms, too, have enzymes with optimum temperatures suited to their environments. Some of them are able to live in extremely hot conditions, and enzymes with optimum temperatures of 80 °C or even higher are not uncommon.

Metabolic reactions in an organism are almost all catalysed by enzymes, so rate of metabolism is affected by temperature in a similar way to that shown in Figure 15.

This in turn affects the rate of growth and development of cells, and therefore the rate of development of the whole organism.

Investigating the effect of temperature on brine shrimp hatch rates

Brine shrimp, *Artemia* sp., are small crustaceans that live in water where the salt concentration is high, such as in salt lakes. These lakes often dry out. Brine shrimp eggs are able to survive for long periods in dry conditions, and hatch within a few days when they are covered with salt water again. You can get brine shrimp eggs from shops selling aquarium supplies.

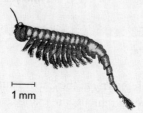

1 mm

Figure 16 Brine shrimp

Place equal volumes of salt water (preferably using a sea water mix for a marine aquarium, or mix approximately 35 g of sodium chloride into 1 dm^3 of distilled water) into several containers. Place each container in a different temperature. You should use at least six different temperatures, covering a range of just over 0 °C to about 80 °C. Keep all other conditions, such as light intensity and salinity of the water, constant. Leave the containers until the water has come to the appropriate temperature, and maintain the temperature at this level.

Add equal quantities of brine shrimp eggs to each container. Check at intervals for the presence of tiny swimming larvae (called nauplii). You should try to check at regular intervals over a 72 hour period, so you will need to share the work out. You may need to use a lens to see the eggs and larvae clearly, or you can take samples using a wide-mouthed pipette and place in a Petri dish under a binocular microscope to observe them. Alternatively, they can be placed in a cavity slide, covered with a coverslip and observed with an ordinary microscope. Record the time at which hatching begins in each beaker.

Evidence for global warming

Records of carbon dioxide levels

The graphs in Figure 11 on page 19 show how carbon dioxide levels in the atmosphere have changed over the last 10 000 years. Data for the earliest years have been obtained by analysing the bubbles of gas trapped in ice. Drills are used to cut deep down into thick layers of ice in the Arctic ice sheets, and bring ice cores to the surface. The deepest parts of these ice cores formed thousands of years ago.

More recently, the atmospheric carbon dioxide levels have been measured regularly at Mauna Loa, Hawaii. These records show how carbon dioxide levels fluctuate with the seasons, as well as longer-term trends.

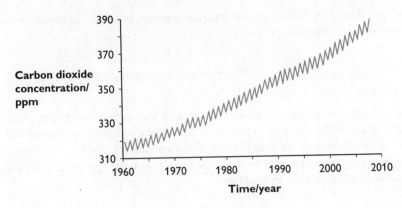

Figure 17 Mean carbon dioxide concentration measured at Mauna Loa

Knowledge check II

Suggest why carbon dioxide concentration goes up and down each year.

Temperature records

Temperatures are monitored regularly at the Earth's surface at many different points in the world. Many of these measurements are made remotely by satellites. The graph in Figure 13 on page 20 shows how mean surface temperature has changed since 1860.

Pollen in peat bogs

Each species of flowering plant has distinctively shaped pollen grains. These survive for long periods of time in peat bogs, where there is not enough oxygen to allow decay organisms to thrive. By analysing the pollen grains found at different levels in peat deposits, we can identify the species of plants that lived there in the past.

We know the climatic conditions in which particular species are able to live, so their presence or absence at a particular time and place can give us information about the climate. For example, beech pollen indicates a warm climate, while fir and pine trees indicate a cooler climate. Pollen evidence from temperate regions in the northern hemisphere shows that there have been periods of warming and cooling in the past 300 000 years, and that current temperatures are relatively high.

Dendrochronology

Dendochronology is the study of tree rings. Each year, a tree makes new xylem vessels in its trunk and branches, and these remain visible as rings. When growth conditions are good (which for most tree species means warm temperatures and plenty of water), rings are broad.

As there is one ring per year, it is possible to count exactly how long ago each ring was formed. By looking at many trees and averaging their growth rates in particular years, we can deduce the climatic conditions at different periods.

We can even do this for trees that were cut down at an unknown time. This is because the relative sizes of the rings in all the different trees in an area will show the same pattern over the same period of time. So if we know the exact age of one tree, we can match up the pattern of rings against another, older, tree, to find a period of overlap. By doing this for many trees, we can match up even very old pieces of wood to a particular time period.

Edexcel A2 Biology

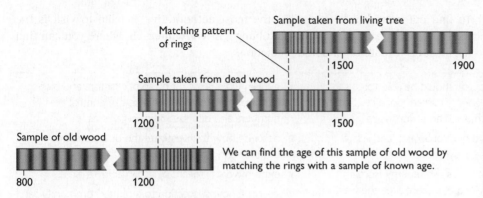

Figure 18 Dating using tree rings

Knowledge check 12

Use the diagrams to estimate when the tree, from which the old wood was taken, was cut down.

Predictions of future climate change

It is not possible to predict exactly what is going to happen to concentrations of greenhouse gases, or to global temperatures, in the future. By collecting together the data we have about changes happening now, and their probable causes, models can be constructed that can make predictions about the range of possible temperature rises we might expect in the next decade or next century. However, as we do not have an absolutely clear understanding of the different processes that are happening now, it is not possible to know exactly what will happen in the future. For example:

- We know that levels of greenhouse gases and mean global temperature are both rising, but we cannot be sure that the temperature rises are entirely a result of the greenhouse gas increases — there could be some other factor that is also contributing.
- We do not know by how much carbon dioxide and methane emissions will increase in the next few years, because this depends on actions taken by many different countries all over the world.
- We know that tropical rainforests take large amounts of carbon dioxide from the air, but we do not know exactly how much they absorb, nor can we predict how they will respond to increasing carbon dioxide levels, especially as these will affect temperature, humidity and rainfall and therefore the rate of growth of the trees.

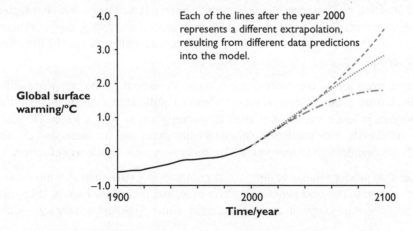

Knowledge check 13

Suggest why the line begins at −0.6 °C, not 0 °C.

Figure 19 Predictions of global temperature change by 2100

To find out more about this topic, the most authoritative website to visit is the Intergovernmental Panel on Climate Change, at **www.ipcc.ch**, where you can find the most up-to-date reports and predictions.

Summary

After studying this topic, you should be able to:
- describe ATP as the energy currency of a cell, and outline how it is synthesised and hydrolysed
- describe the structure of chloroplasts, and relate this to their role in photosynthesis
- outline how photosynthesis uses energy from light to split water, make glucose and produce the waste product oxygen
- describe the light-dependent reactions of photosynthesis, and how these generate reduced NADP and ATP
- describe the light-independent reactions of photosynthesis, and how these use the products of the light-dependent reactions to synthesise carbohydrates
- explain what is meant by gross primary productivity, net primary productivity and efficiency of energy transfer, and carry out calculations involving these
- explain what is meant by abiotic factors and biotic factors, and explain how these can control the numbers and distribution of organisms
- describe how to investigate the distribution of organisms in a habitat
- explain what is meant by an organism's niche
- describe how succession to a climax community can take place
- describe the carbon cycle, and discuss how human activities are contributing to global warming
- discuss the effects of global warming on plants and animals
- describe how to investigate the effects of temperature on brine shrimp
- analyse and discuss the evidence for global warming, and how data can be used to make predictions for the future and to find ways of reducing global warming

Natural selection and evolution

We have seen that each species of organism occupies a particular niche in an ecosystem, and that this niche is related to the adaptations of that species which enable it to survive when affected by a particular set of abiotic and biotic factors. These adaptations arise through mutation and natural selection.

You already know that many of an organism's characteristics are determined, or at least affected, by its genes. Some of these characteristics may also be affected by its environment, but this does not usually have any effect on its genes and therefore these environmentally-caused changes in its characteristics cannot be inherited by its offspring.

In any population of organisms, genes come in various forms, known as alleles. Mutation may produce new alleles. Different combinations of alleles in different individuals produce variation in their characteristics. In your AS studies, you saw that individuals with particular characteristics may have an increased chance of surviving long enough to reproduce. This process is called **natural selection**.

Alleles that produce these advantageous characteristics are therefore more likely to be passed on to the next generation. Over time, and many generations, these alleles will become more common in the population, while other less advantageous alleles become less common. **Evolution** happens when there is a change in allele frequency over time.

Usually, a species is already well adapted to its environment, so allele frequency remains fairly constant over many generations. However, if the environment changes (for example, if the climate becomes warmer), or if a new advantageous allele arises by mutation, then natural selection may produce a change in allele frequency.

Speciation

A species is often defined as a group of organisms with similar morphological and physiological characteristics, which are able to breed with each other to produce fertile offspring. So, for example, lions and tigers are distinct species, even though in a zoo they may be persuaded to breed together. Such interbreeding between the two species never occurs in the wild and, in any case, the offspring are not able to breed themselves.

So how are new species produced? We have seen that natural selection can produce changes in allele frequency in a species, but how much change is needed before we can say that a new species has been formed?

The crucial event that must occur is that one population must become unable to interbreed with another. They must become **reproductively isolated** from one another. Once this has happened, we can say that the two populations are now different species.

There are many ways in which reproductive isolation can happen. One which we think has been especially important in the formation of new species of plants and animals begins by a group of individuals in the population becoming geographically separated from the rest.

For example, a few lizards might get carried out to sea on a floating log, and be carried to an island where that species of lizard was not previously found. This island group is subjected to different environmental conditions from the rest of the species, left behind on the mainland. Different alleles are therefore selected for in the two groups. Over time, the allele frequency in the island group becomes very different from the allele frequency in the original, mainland lizards. This may cause their characteristics to become so different that — even if a bridge appears between the two islands — they can no longer interbreed to produce fertile offspring. Reasons for this could include:

- they have evolved different courtship behaviours, so that mating no longer occurs between them
- the sperm of one group are no longer able to survive in the bodies of the females of the other group, so fertilisation does not occur.
- the number or structure of the chromosomes is different, so that the zygote that is formed by fertilisation does not have a complete double set of genes and cannot develop
- even if a zygote is successfully produced, the resulting offspring may not be able to form gametes, because its two sets of chromosomes (one from each parent) are unable to pair up with each other successfully and so cannot complete meiosis.

Knowledge check 14

Suggest why it is difficult to decide if two fossils belong to the same species or different species.

Examiner tip

Take care to use the terms 'allele' and 'gene' correctly. Alleles are forms of a gene.

Genomes and proteomes

Evolution and speciation take time, and therefore we are not usually able to watch them happening. The evidence that we have provides us with clues about what may have happened in the past. This evidence is often circumstantial, and its correct interpretation depends on a rigorous scientific approach.

Researchers have a wide range of types of evidence they can use to try to work out how evolution has happened in the past. Now that we have a good understanding of DNA and how it affects protein synthesis, studies of DNA and proteins in different species have become a very important tool in investigating evolutionary relationships.

For example, we may want to find out how closely two species of lizard are related to one another. DNA samples can be taken from two species and their base sequences compared. If the species only split apart from one another a short while ago, then we would expect their base sequences to be quite similar. The longer ago they diverged — that is, the less closely related they are — the more differences in base sequence we would expect to find.

Similarly, we can look at the sequences of amino acids in their proteins. Each species has genes that code for the production of a particular set of proteins. The complete set of genes is known as the **genome**, and the complete set of proteins as the **proteome**. The more similar the genomes or the proteomes of two species, the more closely they are thought be related.

Communicating and evaluating the evidence

Evidence for evolution, and about the degree of relationships between species, comes from many different sources, such as the fossil record, similarities in physiology and morphology, and the degree of similarity in the base sequences of DNA and in the amino acid sequences of proteins. If all these different lines of evidence suggest a particular pattern of relationship, we can put quite high trust in our interpretation of the data. If they disagree, we need to collect even more evidence.

Researchers all over the world carry out meticulously planned experiments, collecting large amounts of data which they analyse and interpret. Many of these experiments are published as scientific papers in journals, so that they become available to other researchers who can plan new experiments to test the conclusions even more rigorously, or to build on the new information that has been discovered.

Before a scientific paper is published, it is sent to well-respected researchers in a similar field, who are asked to read it and comment on any problems they can see with the method, the records of the results, the statistical analyses or the conclusions that have been drawn. This is called peer review. Most scientific journals only publish papers after a rigorous peer review has taken place, so that the information in a published paper can generally be trusted.

Researchers also share ideas at conferences held in different parts of the world. A particular research group may be asked to give a talk about their work, and can be questioned about it. This allows researchers working in similar areas of biology to learn what others are doing and to interact and discuss how their findings interrelate.

Summary

After studying this topic, you should be able to:
- explain the meanings of the terms evolution and natural selection
- explain how reproductive isolation may lead to speciation
- discuss how evidence for evolution is communicate and evaluated in the scientific community

Infection, immunity and forensics

Genes and protein synthesis

A DNA molecule is made up of two chains of nucleotides, twisted around each other to form a double helix. Each nucleotide contains one of four bases — adenine, cytosine, guanine or thymine.

Hydrogen bonds link complementary bases

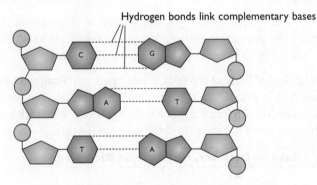

Figure 20 Part of a DNA molecule

Knowledge check 15

Use the diagram to explain why A can bond only with T, and not with another A, or with C or G.

The sequence of bases in one of the DNA strands (the template strand) determines the sequence of amino acids in a protein made by the cell. A length of DNA that codes for one polypeptide or one protein is called a **gene**. A group of three bases, known as a **triplet**, codes for one amino acid.

Table 1 The genetic code

1st base		2nd base							
		A		**G**		**T**		**C**	
A		AAA	Phe	AGA	Ser	ATA	Tyr	ACA	Cys
		AAG	Phe	AGG	Ser	ATG	Tyr	ACG	Cys
		AAT	Leu	AGT	Ser	ATT	Stop	ACT	Stop
		AAC	Leu	AGC	Ser	ATC	Stop	ACC	Trp
G		GAA	Leu	GGA	Pro	GTA	His	GCA	Arg
		GAG	Leu	GGG	Pro	GTG	His	GCG	Arg
		GAT	Leu	GGT	Pro	GTT	Gln	GCT	Arg
		GAC	Leu	GGC	Pro	GTC	Gln	GCC	Arg
T		TAA	Ile	TGA	Thr	TTA	Asn	TCA	Ser
		TAG	Ile	TGG	Thr	TTG	Asn	TCG	Ser
		TAT	Ile	TGT	Thr	TTT	Lys	TCT	Arg
		TAC	Met	TGC	Thr	TTC	Lys	TCC	Arg
C		CAA	Val	CGA	Ala	CTA	Asp	CCA	Gly
		CAG	Val	CGG	Ala	CTG	Asp	CCG	Gly
		CAT	Val	CGT	Ala	CTT	Glu	CCCT	Gly
		CAC	Val	CGC	Ala	CTC	Glu	CCC	Gly

You can see that each triplet codes for only one amino acid, so the code is said to be **non-overlapping**. However, there is more than one triplet coding for most amino acids, so the code is said to be **degenerate**.

Transcription

The first step in protein synthesis is for the sequence of bases on the template strand of the DNA to be used to construct a strand of messenger RNA (mRNA) with a complementary sequence of bases. This is called **transcription**.

In the nucleus, the double helix of the DNA is unzipped, exposing the bases on each strand. There are four types of free RNA nucleotide in the nucleus, with the bases A, C, G and U (uracil). The RNA nucleotides form hydrogen bonds with the exposed bases on the template strand of the DNA. They pair up like this:

Base on DNA strand	Base on RNA strand
A	U
C	G
G	C
T	A

As the RNA nucleotides slot into place next to their complementary bases on the DNA, the enzyme **RNA polymerase** links them together (through their sugar and phosphate groups) to form a long chain of RNA nucleotides. This is an mRNA molecule.

The mRNA molecule contains a complementary copy of the base sequence on the template strand of part of a DNA molecule. Each triplet on the DNA is represented by a complementary group of three bases on the mRNA, called a **codon**.

1 Part of a molecule of DNA

2 The hydrogen bonds between bases are broken, exposing the bases

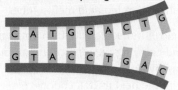

Knowledge check 16

How many amino acids are coded for by this length of DNA?

3 Free RNA nucleotides in the nucleus form new hydrogen bonds with the exposed bases on the template strand

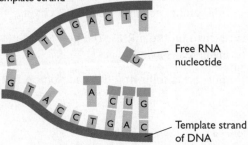

Free RNA nucleotide

Template strand of DNA

4 The RNA nucleotides are linked together to form an mRNA molecule

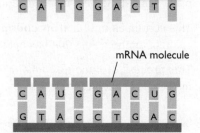

mRNA molecule

Figure 21 Transcription of part of a DNA molecule

Translation

The mRNA molecule breaks away from the DNA, and moves out of the nucleus into the cytoplasm. It becomes attached to a ribosome. Two codons fit into a groove in the ribosome. The first codon is generally AUG, which is known as a **start codon**. It codes for the amino acid methionine.

In the cytoplasm, 20 different types of amino acids are present. There are also many different types of transfer RNA (tRNA) molecules. Each tRNA molecule is made up of a single strand of RNA nucleotides, twisted round on itself to form a clover-leaf shape. There is a group of three exposed bases, called an **anticodon**. There is also a position at which a particular amino acid can be loaded by a specific enzyme.

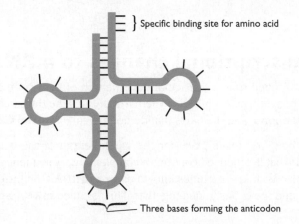

} Specific binding site for amino acid

Three bases forming the anticodon

Figure 22 A tRNA molecule

Knowledge check 17

What will the tRNA anticodons be, to fit against the mRNA shown in Figure 21?

The amino acid that can be loaded onto the tRNA is determined by the base sequence of its anticodon. Thus, for example, a tRNA whose anticodon is UAC will be loaded with the amino acid methionine.

A tRNA molecule with the complementary anticodon to the first codon on the mRNA, and carrying its appropriate amino acid, slots into place next to it in the ribosome, and hydrogen bonds form between the bases. Then a second tRNA does the same with the next mRNA codon.

The amino acids carried by the two adjacent tRNAs are then linked by a peptide bond.

The mRNA is then moved along one place in the ribosome, and a third tRNA slots into place against the next mRNA codon. A third amino acid is added to the chain.

This continues until a **stop codon** is reached on the mRNA. The polypeptide (long chain of amino acids) that has been formed breaks away.

This process of building a chain of amino acids following the code on an mRNA molecule is called **translation**.

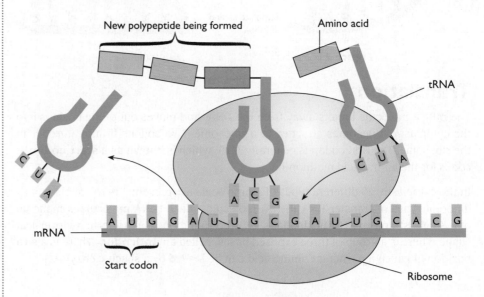

Figure 23 Translation

Post-transcriptional changes to mRNA

The mRNA transcribed from a particular gene (length of DNA coding for a protein) is not all used to synthesise a protein. The parts of the gene that do code for amino acids are called **exons**, and the parts that do not are called **introns**.

Transcription, however, takes place for the whole length of the gene, so the first mRNA molecule that is produced contains a complementary sequence for the introns as well as the exons. It is sometimes known as **pre-mRNA**. The intron sections are then cut away, and the exons joined together. This is called **mRNA splicing**.

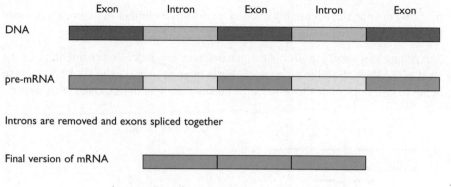

Figure 24 Post-transcriptional mRNA splicing

The final verson of the mRNA molecule, now containing only the code from the exons, leaves the nucleus and travels to a ribosome.

The pre-mRNA molecules can be spliced together in different ways. This means that one gene can produce several different versions of mRNA, which, of course, means that different sequences of amino acids are coded for. Thus one gene can code for several different proteins.

DNA profiling

We have seen that differences and similarities in the base sequences in the DNA of different species can be used to provide one line of evidence for the degree of relationship between them. Characterising the base sequence in the DNA of an organism is called **DNA profiling**.

Even within a species, different individuals have differences in the base sequence of some regions of their DNA. Particular regions of the genome are especially variable, and the base sequences in some of these regions are only likely to be identical in a tiny number of cases. DNA samples found at the scene of a crime can be sequenced, and matched against the sequences of people known to be at the scene and also any suspects. If the sequences of two samples match perfectly, there is a very high chance that they came from the same person.

DNA profiling can also be used to determine the father of a child. The base sequences on the child's DNA must have come from its mother or its father. If a particular base sequence in the child's DNA does not match its mother's, then this base sequence will be found in the father's DNA. If a man thought to be the father does not have this base sequence, then he is not the biological father of the child (see Figure 26 on page 35).

The polymerase chain reaction

The polymerase chain reaction is usually known as PCR. It is an automated method of making multiple copies of a tiny sample of DNA. This enables DNA profiling to be carried out, even if only a very small amount of DNA was originally available.

PCR involves exposure of the DNA to a repeating sequence of different temperatures, allowing different enzymes to work. The process shown in the diagram is repeated

over and over again, eventually making a very large number of identical copies of the original DNA molecule.

The primer is a short length of DNA with a base sequence complementary to the start of the DNA strand to be copied. This is needed to make the DNA polymerase begin to link nucleotides together as it makes a copy of the exposed DNA strand.

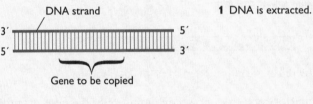

1 DNA is extracted.

Knowledge check 18

Suggest the purpose of the primer used in PCR.

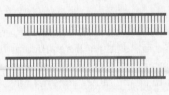

2 The DNA is heated to 95 °C to denature it, which separates the double helix.

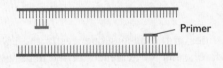

3 The DNA is cooled to 65 °C, and primer DNA is added.

Complementary base pairing occurs.

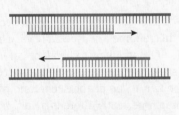

4 The DNA and primers are incubated at 72 °C with DNA polymerase and free nucleotides to synthesise complementary strands of DNA.

5 The DNA has been copied, and each of the two new strands form part of two DNA molecules.

6 The DNA is heated to 95 °C again to denature the DNA, and a new cycle of copying occurs, following steps 2–5.

This is repeated many times to synthesise many new copies of DNA.

Figure 25 The polymerase chain reaction

Gel electrophoresis

Gel electrophoresis is a way of separating strands of DNA of different lengths.

Before electrophoresis is carried out, a sample of DNA is exposed to a set of **restriction enzymes**. These enzymes cut DNA molecules where particular base sequences are present. For example, a restriction enzyme called *Bam*H1 cuts where the base sequence -GGATCC- is present on one strand of the DNA. Other restriction enzymes target different base sequences. This cuts the DNA into fragments of different lengths.

Gel electrophoresis can then be used to separate out these different lengths of DNA. If two samples of DNA are treated with the same restriction enzymes, and the fragment lengths are compared, this can help to decide whether the DNA samples came from the same individual or not.

To carry out gel electrophoresis, a small, shallow tank is partly filled with a layer of agarose gel. A potential difference is applied across the gel, so that a direct current flows through it.

A mix of the DNA fragments to be separated is placed on the gel. DNA fragments carry a small negative charge, so they slowly move towards the positive terminal. The larger they are, the more slowly they move. After some time, the current is switched off and the DNA fragments stop moving through the gel.

The DNA fragments must be made visible in some way, so that their final positions can be determined. This can be done by adding fluorescent markers to the fragments. Alternatively, single strands of DNA made using radioactive isotopes, and with base sequences thought to be similar to those in the DNA fragments, can be added to the gel. They will pair up with fragments which have complementary base sequences, so their positions are now emitting radiation. This can be detected by its effects on a photographic plate.

The dark areas on these autoradiographs of DNA represent particular DNA sequences.

In the first profile, you can see that all the bands on the child's results match up with either the mother's or potential father's, so this man could be the child's father.

However, on the second one, the child has a band that is not present in either the mother's or father's results. Some other person must therefore be the child's father.

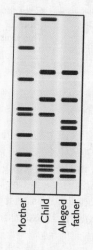
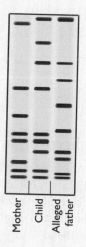

Figure 26 Electrophoresis in paternity testing

Summary

After studying this topic, you should be able to:
- describe the genetic code
- describe how transcription takes place
- describe how translation takes place
- explain how post-transcriptional changes to mRNA can result in several different proteins arising from one gene
- explain how DNA profiling is carried out
- describe the polymerase chain reaction and gel electrophoresis, and their uses

Infection and immunity

Micro-organisms

Figure 27 shows the structure of a bacterium and a virus. These are two very different types of micro-organism. Viruses are many times smaller than bacteria, and are not made of cells. Bacteria are prokaryotes, whose cells do not contain a membrane-bound nucleus.

> **Knowledge check 19**
>
> Name two structures found in an animal cell that are not present in a bacterium.

Bacterium

- Cell wall made of peptidoglycans
- Plasma (cell surface) membrane
- Cytoplasm
- Circular DNA
- Ribosomes (20 nm diameter)

0.5 μm

Virus

- Capsid made up of protein subunits
- Core containing polynucleotide (RNA or DNA). This virus contains RNA.

50 nm

Figure 27 Structures of a bacterium and virus

Micro-organisms in nutrient cycles

Micro-organisms play an essential role in the recycling of carbon (Figure 10, page 19). Many bacteria and fungi are decomposers, secreting hydrolytic enzymes to break down biological material into soluble compounds that are then available to other organisms. Their respiration releases carbon dioxide from organic molecules back to the air.

Edexcel A2 Biology

Micro-organisms and disease

Some micro-organisms, especially bacteria, viruses and a few fungi, cause disease in animals and plants. They are known as **pathogens**.

To cause disease in humans, pathogens must enter the body. They can do this in several ways.

- Through the **skin**. Most micro-organisms cannot penetrate unbroken skin, although a few — such as the virus that causes warts — can do so. We have our own population of harmless bacteria on the skin, some of which produce an environment that is not suitable for the growth of other, potentially harmful, micro-organisms. Skin is covered with a layer of dead, keratinised cells which is difficult to penetrate. Cuts are quickly sealed by blood clotting, preventing entry of pathogens. Eyes are protected by tears containing lysozyme, an enzyme that kills bacteria.

- Through the **alimentary canal**. Pathogens can enter the body in food or drink, or through mouth contact with dirty hands or other objects. We have natural defences against this, such as not wanting to eat food that looks or smells 'bad', or vomiting after eating something distasteful. Hydrochloric acid in the stomach kills many micro-organisms. As on the skin, we have our own natural 'flora' of bacteria inhabiting the alimentary canal whose presence limits the ability of other, harmful, micro-organisms to grow.

- Through the **gas exchange system**. Several types of bacteria and viruses can enter the body as tiny droplets in the air we breathe in, and infect cells in the respiratory passages and the lungs. Mucus produced by goblet cells traps many of these before they reach the lungs, and cilia move the mucus up to the back of the throat where it is swallowed. Phagocytic white blood cells patrol the lungs and destroy pathogens.

- Through the **reproductive tract**. Several bacteria and viruses can be passed from one person to another in fluids produced in the reproductive system. We have few natural defences against these, apart from codes of behaviour that limit the number of sexual partners a person has.

- By **vectors**. A vector in the biological sense is an organism that transmits a pathogen from one host to another. Mosquitoes, for example, are vectors for the pathogen that causes malaria. Their saliva contains the parasites, which are injected into a person's body when a female mosquito bites in order to feed on blood.

Examiner tip

A vector does not *cause* a disease — it just transmits it.

Tuberculosis

Tuberculosis (TB) is caused by a bacterium called *Mycobacterium tuberculosis*. This pathogen enters the body in droplets in inspired air and infects cells in the lungs. Certain white blood cells ingest the bacteria by phagocytosis, but are not able to destroy them. The bacteria remain inside the cells where they may divide, releasing more bacteria that are taken up by other cells.

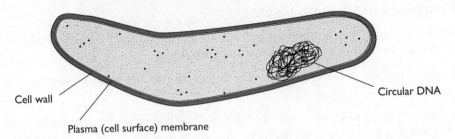

Figure 28 Mycobacterium tuberculosis

Many infected people show no symptoms, but when the population of bacteria becomes large, substantial areas of tissue become damaged. The person loses weight and develops a cough which, as the disease progresses, begins to bring up blood from the lungs. Night sweats are common. If TB is untreated, death may occur.

HIV/AIDS

AIDS is caused by the human immunodeficiency virus, **HIV**. The virus enters the body in the body fluids of an infected individual. It may be passed on during sexual intercourse, or through contaminated blood. It can also be passed from a mother to her unborn child.

The RNA of the HIV enters a particular type of white blood cell called a **T4 cell**. A viral enzyme, **reverse transcriptase**, also enters the cell. This enzyme uses the viral RNA to produce a complementary strand of DNA, which can be used as the code for synthesising new viral proteins. Many new viruses are synthesised in the cell, and eventually burst out and destroy it.

Usually, the virus 'lies low' in the infected cell for some time — often many years — before multiple copies of the virus are made, or large numbers of T cells are destroyed. Symptoms therefore do not appear immediately.

T cells are an important part of the immune system (see below) so if they are put out of action by HIV then the body becomes unable to resist infection by other pathogens, or to destroy its own cells if these have begun to malfunction (for example, if they have become cancerous). The symptoms of HIV are caused by opportunistic infections caused by these pathogens, such as TB, or by the development of unusual cancers, such as Kaposi's sarcoma. If AIDS is left untreated, death almost invariably occurs.

Response of the body to infection

Non-specific responses

When infection (entry of a pathogen into the body) occurs, several defence mechanisms are rapidly brought into play.

Inflammation

Damage to body tissues, or the presence of a population of bacteria, causes arterioles in the area to dilate. This increases the blood supply, bringing more white blood cells. Capillary walls become more porous, allowing blood plasma and white cells to seep

Knowledge check 20

Suggest why the enzyme reverse transcriptase was given this name.

out into the surrounding tissues. Some of these white blood cells are phagocytes, and they ingest and destroy pathogens and also damaged tissue.

Lysozyme action

Many body fluids — including saliva and tears — contain the enzyme lysozyme. This enzyme catalyses the hydrolysis of peptidoglycan molecules in bacterial cell walls. Lysozyme is also produced by neutrophils, a type of white blood cell.

Interferon

Interferons are a class of substances that inhibit the replication of viral RNA. Interferons are glycoproteins, and can be produced by many different kinds of body cells when double-stranded RNA (indicating the presence of viruses) is detected. The presence of interferon also triggers the production of various pathogen-fighting substances, and increases the activity of T cells, which destroy body cells that contain viruses.

Phagocytosis

Two types of white blood cell, monocytes and macrophages, are able to ingest and destroy bacteria or other foreign material, including any of the body's own cells that are damaged irretrievably.

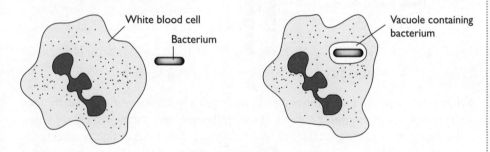

White blood cell

Bacterium

Vacuole containing bacterium

Figure 29 Phagocytosis

The bacterium is taken into a vacuole, sometimes called a phagosome. Lysosomes then fuse with the phagosome, releasing hydrolytic enzymes that break down the bacterium.

The immune response

The responses to infection described above can all happen fairly rapidly after the entry of a pathogen into the body. However, they are all just general responses. We also have another line of defence, in which the responses are specifically targeted against a particular pathogen. This response takes longer to get started, but once under way it is a highly effective method of attack.

The immune response depends on the presence of **antigens** on the invading pathogens. These are molecules, generally on the surface of the pathogen, that are specific to that particular micro-organism. They are frequently proteins or glycoproteins.

Lymphocytes are a group of white blood cells that play a central role in the immune response. There are two main types — **B cells** and **T cells**. They look identical when viewed with a microscope, but they have different functions.

B cells

All our cells contain genes that are able to produce a very wide range of proteins called **immunoglobulins**. These can act as **antibodies**. We have thousands of different B cells, in each of which a different type of antibody can be produced. This is possible because of post-transcriptional processing of the genes that code for immunoglobulins, producing many different versions of mRNA and therefore many different versions of the proteins.

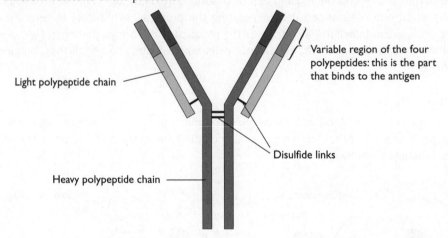

Light polypeptide chain

Variable region of the four polypeptides: this is the part that binds to the antigen

Disulfide links

Heavy polypeptide chain

Figure 30 An immunoglobulin molecule (an antibody)

Examiner tip

Take care not to confuse antigens and antibodies.

B cells place some of these antibodies in their cell surface membranes, where they act as receptors. The variable regions of the antibodies have a complementary shape to the antigens on pathogens against which they act. When this antigen contacts the B cell receptor, the B cell is activated. It divides repeatedly by mitosis, producing a clone of cells.

Some of these develop into **plasma cells**, sometimes called **B effector cells**, which synthesise and secrete large amounts of the antibody. Their activity is so intense that they do not usually live more than a few weeks, but they are replaced by more plasma cells if need be. Others remain in the blood but do not secrete antibody. They are **memory cells**, and they live for a very long time. Their continued presence in the blood, perhaps many years after the original infection, means that the immune system can mount an instant attack on the same pathogen should it invade the body again. The person has become immune to that particular disease.

B cells may intially come into contact with the antigen in the blood plasma or body fluids, or they may meet it on an **antigen-presenting cell**. Several different types of cell, including macrophages, act as antigen-presenting cells. They place antigens of pathogens they have encountered in their cell surface membranes, where there is a good chance that a B cell will encounter them.

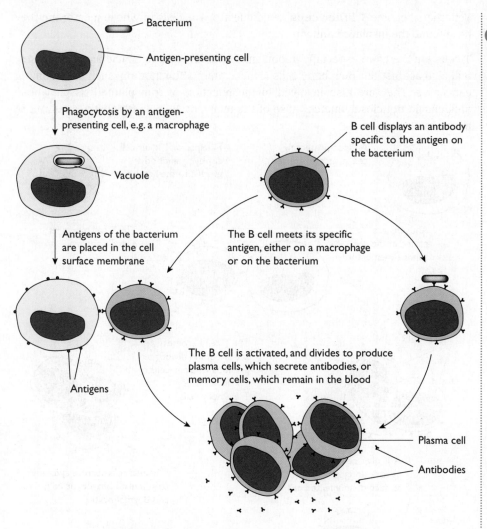

Figure 31 The response of B cells to an antigen

Labels in figure: Bacterium; Antigen-presenting cell; Phagocytosis by an antigen-presenting cell, e.g. a macrophage; Vacuole; Antigens of the bacterium are placed in the cell surface membrane; Antigens; B cell displays an antibody specific to the antigen on the bacterium; The B cell meets its specific antigen, either on a macrophage or on the bacterium; The B cell is activated, and divides to produce plasma cells, which secrete antibodies, or memory cells, which remain in the blood; Plasma cell; Antibodies

T cells

T cells, like B cells, place glycoproteins in their cell surface membranes, and these bind specifically with antigens. However, T cells will only respond to antigens if they find them in the cell surface membranes of body cells. These may be antigen-presenting cells such as macrophages, or they may be cells that have been infected by viruses and have placed molecules from the virus in their membranes as a 'help' signal.

T cells respond in a similar way to B cells when they meet their antigen, quickly producing a clone. Some of these remain inactive, as **T memory cells**, lasting for many years. Others become **T helper cells**, which secrete **cytokines** (for example interferon) that stimulate other cells, such as macrophages, to become active against the virus.

Yet others become **T killer cells**, which destroy the cells in whose membrane they have found the displayed antigen.

T cells are therefore especially important in the immune response against viruses, and also against our own body cells if these are malfunctioning, such as becoming cancerous. They are also involved in the rejection of transplanted tissue whose antigens do not closely match those of the recipient. It is T cells that are put out of action by HIV.

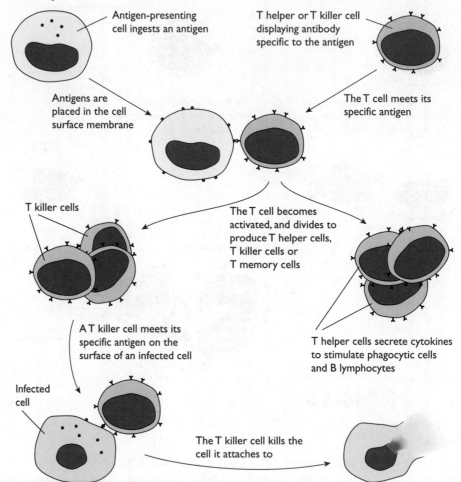

Figure 32 The response of T cells to antigen

Antigen-presenting cell ingests an antigen

T helper or T killer cell displaying antibody specific to the antigen

Antigens are placed in the cell surface membrane

The T cell meets its specific antigen

T killer cells

The T cell becomes activated, and divides to produce T helper cells, T killer cells or T memory cells

A T killer cell meets its specific antigen on the surface of an infected cell

T helper cells secrete cytokines to stimulate phagocytic cells and B lymphocytes

Infected cell

The T killer cell kills the cell it attaches to

> **Examiner tip**
> Do not confuse cytokines (produced by T cells) with antibodies (produced by B cells).

The development of immunity

Active immunity

This occurs when the person has made their own antibodies against a pathogen, either as the result of an infection as described above (natural immunity) or after being vaccinated with a weakened form of the pathogen (artificial immunity). Memory cells remain in the body for many years, so this type of immunity tends to be very long-lasting.

Passive immunity

This occurs when ready-made antibodies enter the person's body. A fetus acquires antibodies from its mother while in the uterus, and there are also antibodies in breast milk. This is a form of natural immunity.

A person can also obtain antibodies by injection. This may be done if they are thought to have already been infected with a serious disease, and need instant help to destroy the pathogens. Antibodies may also be given to people travelling to an area where there is a high risk of infection, such as aid workers going to earthquake zones. This is a form of artificial immunity.

Passive immunity generally lasts only weeks or months at best, because the person does not have their own memory cells, and the antibodies they have been given do not last long.

The evolutionary arms race

Pathogens that can defeat a person's immune system have a better chance of surviving and reproducing than those that cannot. There is strong natural selection for individual pathogens that possess features enabling them to evade attack. At the same time, there is strong natural selection for individual hosts that are able to resist the invasions of pathogens. As a new evasion mechanism develops in a pathogen, so selection pressures on the host change, favouring individuals who happen to have alleles that provide them with new ways of defeating the pathogens.

HIV, for example, causes the T cells that it has invaded to fail to place antigens in their cell surface membranes. They are therefore not attacked by other T cells which would normally respond to viral infection by destroying them. This effect appears to be caused by a protein encoded by the viral RNA, called Nef. The production of this protein has evolved as the result of natural selection against viruses that do not produce it.

Immune system evasion mechanisms are also seen in pathogenic bacteria, such as *Mycobacterium tuberculosis*. This, unusually for bacteria, gets *inside* macrophage cells and breeds there. Being inside the cells, it avoids exposure to B cells, which therefore do not respond to the *M. tuberculosis* antigens by producing antibodies against them. Normally macrophages would produce phagosomes and destroy any bacteria they take up, but *M. tuberculosis* produces substances which inhibit this process so that they can live and breed inside the macrophage indefinitely.

Studies of HIV, *Mycobacterium* and other pathogens show that changes take place in them over time, which appear to be the result of natural selection operating on the natural variation that occurs within their populations. Thus, HIV with the Nef protein has a selective advantage, while HIV without Nef is unlikely to survive in a host.

At the same time, the hosts of these pathogens are also undergoing natural selection. For example, a person whose T cells are not affected by Nef would be able to destroy any HIV in their body and not get AIDS.

Examiner tip

Avoid saying that bacteria 'evolve so that they become' able to resist the immune response, as this implies they change purposefully.

Antibiotics

Despite the efficiency of the immune system, we do not always win the battle against bacterial pathogens. Infection of wounds by bacteria used to be a major cause of death. Today, we have antibiotic drugs that can be used to kill bacteria inside the body.

Antibiotics are chemicals that kill or weaken bacteria but do not normally harm human cells. They do not kill viruses.

Many antibiotics, including penicillin and vancomycin, damage bacterial cell walls. For example, penicillin inhibits a group of enzymes called glycopeptidases, which build cross-links between the peptidoglycan molecules that make up the cell wall. The cell walls become leaky and the cells burst. These antibiotics do not harm human cells because they do not have cell walls.

Other antibiotics target other metabolic pathways in the bacteria. Tetracycline and erythromycin enter bacterial cells and bind with enzymes involved in protein synthesis on the bacterial ribosomes. Protein synthesis cannot take place, so the bacterial cells are unable to grow and divide. These antibiotics do not harm human cells because our ribosomes and protein-synthesising enzymes are different from those of bacteria.

Antibiotics are sometimes classified as being **bacteriocidal** or **bacteriostatic**. Bacteriocidal antibiotics kill bacteria, whereas bacteriostatic ones slow down or prevent their growth and reproduction. Generally, penicillin and other cell-wall inhibitors are considered to be bacteriocidal, while tetracycline and erythromycin are considered to be bacteriostatic.

However, their effects may differ against different bacteria, or when given in different doses. For example, erythromycin is bacteriocidal against *Streptococcus pneumoniae* but bacteriostatic against *Staphylococcus aureus*.

Knowledge check 21

Explain the difference between an antibody and an antibiotic.

Investigating the effect of antibiotics on bacteria

Throughout this investigation, you need to use sterile technique. This involves taking great care not to allow bacteria from any other source than the intended one to get onto your culture medium. You also need to take care that none of the bacteria you are using get onto you or your clothes.

- Wipe down the bench with disinfectant. Collect all the equipment you need, ensuring that everything is clean. Wear a lab coat and goggles.
- You will need a sterile Petri dish containing sterile agar jelly. Your teacher may provide you with this, or you may be able to prepare the agar yourself and pour it into the dish.
- You also need a culture of harmless bacteria, such as *Escherichia coli*. It is important that the bacteria come from a known source (a biological supplier), so that you are certain they are not pathogenic. The bacteria should be growing in a liquid medium.

- Use a sterile pipette or syringe to take up about 3 cm³ of the bacterial culture. Hold the lid of the Petri dish partly open with one hand, and place the culture on the surface of the agar. Quickly close the lid and gently tip the dish so that the bacterial culture spreads evenly across the surface.
- Now use sterile forceps to take up a disc that has been impregnated with an antibiotic. Place the disc on the surface of the agar, again taking care not to open the lid of the dish too wide or for too long. Repeat with other discs containing different antibiotics (or different concentrations of the same antibiotic). (If the discs are not labelled, you will need to write labels on the Petri dish itself.)
- Incubate the dish at about 27 °C.
- You should see results after a day or so. The individual bacteria in the culture medium, which were too small to see, will have reproduced to form visible colonies on the surface of the agar. If any of the antibiotics were active against the bacteria, there will be clear areas around the discs where the bacteria were unable to grow. By comparing the diameters of these clear areas, you can compare the effectiveness of the antibiotics against the bacteria.
- Immerse the Petri dishes in disinfectant solution to kill all bacteria in them, before disposing of them.

Knowledge check 22

How does the antibiotic spread into the agar?

Hospital-acquired infections

In recent years, there has been great concern about the high rate of infection with pathogenic bacteria among people who have been admitted to hospital for other reasons. Older people are especially susceptible. In some cases, people have died from these infections.

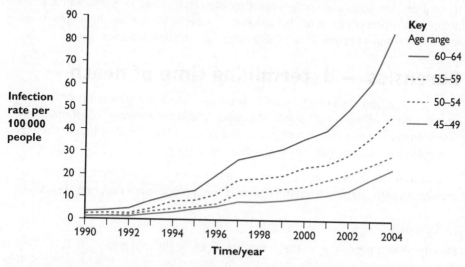

Figure 33 Increase in *Clostridium difficile* infections in hospitals between 1990 and 2004

The reasons for these infections and death include:

- People admitted to hospital are often weak, and their immune systems are not working well. Bacteria that would not normally be able to cause problems, such

as *Clostridium difficile* or *Staphylococcus aureus*, are able to get past their defences and cause serious illness or death.
- Populations of bacteria in hospitals are likely to have been exposed to antibiotics, so that natural selection has resulted in these populations becoming antibiotic-resistant.
- Poor hygiene in hospitals increases the risk of bacteria from one patient being transferred to another.

The National Health Service now follows strict guidelines to attempt to reduce the rates of infection with bacteria in hospitals. For example, patients infected with *C. difficile* often suffer from diarrhoea. Health workers are advised to assume that anyone with diarrhoea may have *C. difficile*, and to reduce the risk of transmitting the bacterium to other patients by following the SIGHT procedure:

Suspect that a case may be infective where is no clear alternative cause for diarrhoea.
Isolate the patient and inform the Infection Control Team.
Gloves and aprons must be worn when contacting the patient or their surroundings.
Hands must be washed with soap and water before and after contacting the patient and their environment.
Tests should be carried out as soon as possible to determine the cause of the diarrhoea.

It is also important to try to reduce the number of populations of bacteria that are becoming resistant to antibiotics. The best way of doing this is to reduce the use of antibiotics. The more they are used, the more bacteria are exposed to them and the greater the selection pressures for antibiotic resistance. GPs and other health workers try to prescribe antibiotics only when they are clearly needed. Some antibiotics are not generally prescribed at all, but are kept in reserve for use against populations of bacteria that have become resistant to other, commonly used ones.

Forensics — determining time of death

We have seen how DNA can be used to help with determining paternity or to identify people who have been at the scene of a crime. These are examples of forensics — the determination of facts for potential use in a court of law. Another important aspect of forensics is working out the time of death of a body.

Many pieces of information can be gathered from a corpse, which can be used together to estimate when death occurred. Some of these are described below.

Decomposition

Shortly after a person dies, their own enzymes begin to digest their cells. This is called **autolysis**. It generally begins in the pancreas, as this is where the highest concentrations of hydrolytic enzymes are found.

Bacteria also begin to break down substances within cells.

These two processes cause gradual changes in the body, including discolouration and the breakdown of tissues. The further advanced these processes are when the body is found, the longer ago death occurred.

Stages of succession

A body can be considered to be an ecosystem. When death first occurs, a particular community of decomposers begins to break the body down. These decomposers cause changes which enable other species to move in. This is an example of succession — gradual changes in a community over time.

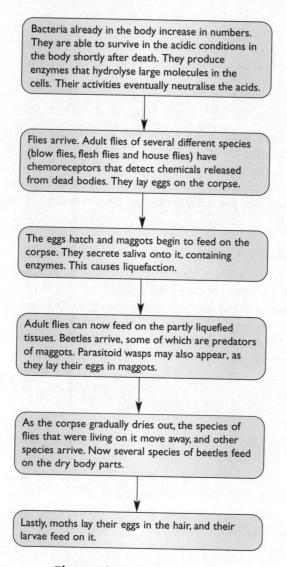

Bacteria already in the body increase in numbers. They are able to survive in the acidic conditions in the body shortly after death. They produce enzymes that hydrolyse large molecules in the cells. Their activities eventually neutralise the acids.

Flies arrive. Adult flies of several different species (blow flies, flesh flies and house flies) have chemoreceptors that detect chemicals released from dead bodies. They lay eggs on the corpse.

The eggs hatch and maggots begin to feed on the corpse. They secrete saliva onto it, containing enzymes. This causes liquefaction.

Adult flies can now feed on the partly liquefied tissues. Beetles arrive, some of which are predators of maggots. Parasitoid wasps may also appear, as they lay their eggs in maggots.

As the corpse gradually dries out, the species of flies that were living on it move away, and other species arrive. Now several species of beetles feed on the dry body parts.

Lastly, moths lay their eggs in the hair, and their larvae feed on it.

Figure 34 Succession in a body

These changes tend always to happen in approximately the same order, but the length of time they last varies according to environmental conditions. Warm, moist conditions will cause more rapid changes than cold, dry ones. The forensic scientist can consult tables which list the times at which you would expect these different communities to be present in different conditions.

Forensic entomology

Entomology is the study of insects. A detailed study of the insects on a corpse can provide evidence about the length of time since death occurred.

You can see from the diagram on page 47 that flies are often the first insects to arrive, followed by beetles and moths. We can also look at the stage of development of a particular species. For example, data have been collected about the rate of development of the eggs and maggots of the fly *Lucilia sericata* at different temperatures. If a forensic scientist finds maggots at a particular stage of development on a corpse, and knows the temperature in which the corpse has been lying, he or she can work out when the eggs were laid.

Body temperature

At death, respiration in cells stops and heat is no longer generated in the body. Core body temperature therefore gradually falls from 37°C to the temperature of the environment. The rate at which this happens depends on the environmental temperature, the amount of adipose (fat) tissue on the corpse and the body mass. Core body temperature has usually reached the environmental temperature by 24 hours after death.

Degree of muscle contraction

Within about 3 hours of death, muscles begin to stiffen. (You will find out why this happens in Unit 5.) This is because the cells are no longer producing ATP. This is called **rigor mortis**. Rigor mortis remains for up to 33 hours. The muscles then become relaxed. Once again, the rate at which rigor mortis sets in and wears away depends on the temperature of the environment, and also various other factors such as how much exercise the person had been doing before death.

Summary

After studying this topic, you should be able to:
- distinguish between the structure of bacteria and viruses
- describe how micro-organisms are involved in nutrient cycles
- describe how pathogens may enter the human body, and the natural defences against this entry
- describe and explain the symptoms caused by infection with *Mycobacterium tuberculosis* and HIV
- describe the non-specific responses of the body to infection
- describe the immune response, including the roles of antigens and antibodies, B cells and T cells

- explain how active immunity and passive immunity may develop
- discuss the 'evolutionary arms race' between pathogens and their hosts, with reference to HIV and *Mycobacterium tuberculosis*
- describe how to investigate the effect of antibiotics on bacteria
- explain the actions that are taken to reduce the incidence of hospital-acquired infections
- describe how time of death of a mammal can be determined

Edexcel A2 Biology

Questions & Answers

In this section there are two sample examination papers, similar to the Edexcel Unit Test papers. All of the questions are based on the topic areas described in the previous sections of the book.

You have 1 hour 30 minutes to do each paper. There are 90 marks on the paper, so you can spend almost 1 minute per mark. If you find you are spending too long on one question, move on to another that you can answer more quickly. If you have time at the end, come back to the difficult one.

Some of the questions require you to recall information that you have learned. Be guided by the number of marks awarded to suggest how much detail you should give in your answer. The more marks there are, the more information you need to give.

Some of the questions require you to use your knowledge and understanding in new situations. Don't be surprised to find something completely new in a question — something you have not seen before. Just think carefully about it, and find something that you do know that will help you to answer it.

Do think carefully before you begin to write. The best answers are short and relevant — if you target your answer well, you can get many marks for a small amount of writing. Don't ramble on and say the same thing several times over, or wander off into answers that have nothing to do with the question. As a general rule, there will be twice as many answer lines as marks. So you should try to answer a 3-mark question in no more than 6 lines of writing. If you are writing much more than that, you almost certainly haven't focused your answer tightly enough.

Look carefully at exactly what each question wants you to do. For example, if it asks you to 'Explain', then you need to say how or why something happens, not just what happens. Many students lose large numbers of marks by not reading the question carefully.

Examiner's comments

Each question is followed by a brief analysis of what to watch out for when answering the question (shown by the icon ⊜). All student responses are then followed by examiner's comments. These are preceded by the icon ⊜ and indicate where credit is due. In the weaker answers, they also point out areas for improvement, specific problems, and common errors such as lack of clarity, weak or non-existent development, irrelevance, misinterpretation of the question and mistaken meanings of terms.

Sample Paper 1

Question 1

The fly *Lucilia sericata* lays its eggs on dead bodies. The rate at which the eggs hatch and develop is partly determined by temperature. The table shows the time at which the different stages of the life cycle appear on a corpse, at three different temperatures.

Temperature/°C	Time since death of body/hours						Total time/days
	Egg	1st instar of larva	2nd instar of larva	3rd instar or larva	Pre-pupa	Pupa	
16	41	53	42	98	148	393	32
21	21	31	26	50	118	240	20
27	18	20	12	40	90	168	14

(a) (i) Calculate the percentage difference in the time taken to reach the 2nd instar stage at 27 °C compared with 21 °C. (2 marks)

(ii) Explain why development happens faster at higher temperatures. (5 marks)

(b) (i) A forensic entomologist found 2nd instar larvae on a corpse found in woodland. Daytime temperatures had been around 18°C for the last week, dropping to 12°C at night. What can the entomologist deduce about the possible time of death? Explain your answer. (3 marks)

(ii) Describe two other types of evidence that a forensic scientist could use to help determine the time of death. (4 marks)

Total: 14 marks

ⓔ Remember to show your working clearly whenever you are asked to do a calculation.

Student A

(a) (i) $12/26 \times 100 = 46\%$ ✗

ⓔ This is incorrect. The student has calculated the 27 °C value as a percentage of the 21 °C value. He or she should have first calculated the difference between the two values, and then calculated this difference as a percentage of the 21 °C value. **0/2**

(ii) Enzymes work faster at higher temperatures because they are moving faster ✓ and have more energy. So metabolism happens faster ✓ and so development happens faster too.

ⓔ This is partly correct, but there is not sufficient detail to get many marks. **2/5**

Edexcel A2 Biology

(b) (i) The temperatures had been close to 16 °C, so the 2nd instar larvae would appear at 42 days. ✓ So the person must have died 42 days ago.

ⓔ The student has used the appropriate row in the table to find relevant data, but he or she has not indicated or explained the considerable degree of uncertainty that the entomologist would have regarding the precise time of death. **1/3**

(ii) Rigor mortis. ✓ This starts about 3 hours after death, and goes by about 33 hours. ✓ Also the species of animals that are found on the body. These show a succession, ✓ so if there are just fly larvae the body has not been there as long as if there are beetles and moth larvae. ✓

ⓔ These answers state two types of evidence and give some detailed information about each one. **4/4**

Student B

(a) (i) 26 − 12 = 14 ✓

$14/26 \times 100 = 54\%$ ✓

The time taken has got shorter, so the percentage change is −54%.

ⓔ An entirely correct answer. **2/2**

(ii) As temperature increases, enzyme and substrate molecules gain kinetic energy, ✓ so they collide more frequently and reaction rate increases. ✓ This is true for all the metabolic reactions ✓ taking place in the fly larvae, and this affects their rate of development. However, if temperatures rose above the optimum temperature ✓ of their enzymes, then the enzymes would begin to denature ✓ and reaction rates would fall, and so would the rate of development.

ⓔ A good answer. The student has clearly explained how temperature affects metabolic reactions, and has linked this to the rate of development. **5/5**

(b) (i) The rates of development at temperatures of 18 °C and 12 °C are not shown in the table. The nearest is 16 °C. ✓ If the body had been at a constant temperature of 16 °C, we would expect to see 2nd instar larvae 42 days after death. ✓ The entomologist could conclude that the person died about 42 days ago, but it could have been several days either side of that. ✓ It must be longer than 26 days ago, as the temperature has never risen as high as 21 °C. ✓

ⓔ A good answer. The student has used the table correctly, and has fully explained the difficulty of drawing any firm conclusions about the time of death. **3/3**

(ii) Body temperature gradually drops ✓ and rigor mortis sets in and then goes. ✓

ⓔ Two appropriate types of evidence have been mentioned, but neither of them has been properly described. **2/4**

Question 2

The diagram shows a stage during protein synthesis.

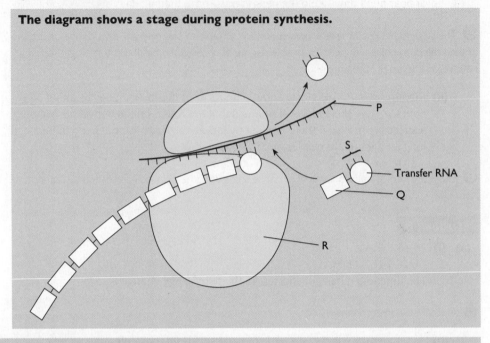

(a) What is the correct name of this stage? Choose from:

polymerase chain reaction   post-transcriptional modification

transcription   translation   (1 mark)

(b) Name the parts labelled P, Q, R and S on the diagram.   (4 marks)

(c) With reference to the diagram, describe the role of tRNA in determining the sequence of amino acids in a protein molecule.   (5 marks)

Total: 10 marks

ⓔ Think about how you can get all 5 marks for part (c) — make sure you have at least 5 strong points in your answer.

Student A

(a) translation ✓

ⓔ Correct. **1/1**

(b) P mRNA, ✓ Q amino acid, ✓ R ribosome, ✓ S codon ✗

ⓔ One mistake. **3/4**

(c) tRNA picks up amino acids in the cytoplasm ✓ and brings them to the ribosome. It binds with the mRNA ✓ and holds the amino acids in place so they can be linked together by peptide bonds. ✓

ⓔ This is correct as far as it goes, but there is no mention of the specificity of each tRNA molecule, and how the complementary base pairing between the tRNA and mRNA ensures that the appropriate amino acid is added to the chain. **3/5**

Student B

(a) translation ✓

ⓔ Correct. **1/1**

(b) P mRNA, ✓ Q amino acid, ✓ R ribosome, ✓ S anticodon ✓

ⓔ All correct. **4/4**

(c) Each tRNA molecule has a particular sequence of bases in its anticodon ✓ and a specific amino acid ✓ it can bind with. The anticodon of the tRNA forms hydrogen bonds with a codon ✓ on the mRNA in the ribosome, by complementary base pairing. ✓ This ensures the amino acid is the one coded for by the mRNA ✓ (and therefore originally by the DNA). Two tRNAs hold two adjacent amino acids in place so they can be joined by a peptide bond. ✓

ⓔ A clear and complete answer. **5/5**

Question 3

The diagram shows the structure of a chloroplast.

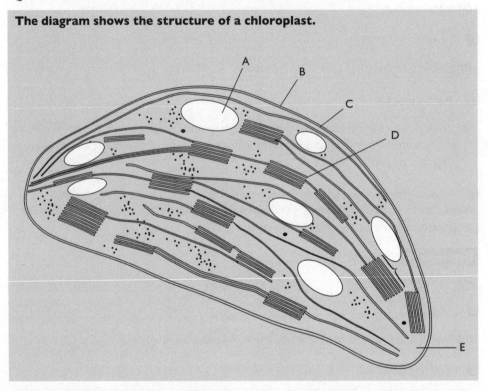

(a) Give the letter of the part of the chloroplast where each of the following takes place.
   (i) fixation of carbon dioxide   (1 mark)
   (ii) the light-dependent reactions   (1 mark)

(b) A grass adapted for growing in a tropical climate was exposed to low temperatures for several days. The membranes of part D moved closer together, so that there was no longer any space between them. This prevented photophosphorylation taking place.
Explain how this would prevent the plant from synthesising carbohydrates.   (4 marks)

(c) Two groups of seedlings were grown in identical conditions for 2 weeks. One group was then grown in high-intensity light and the other group in low-intensity light, for 4 weeks.
Each group of plants was then placed in containers in which carbon dioxide concentration was not a limiting factor. They were exposed to light of varying intensities and their rate of carbon dioxide uptake was measured.
The results are shown in the graph.

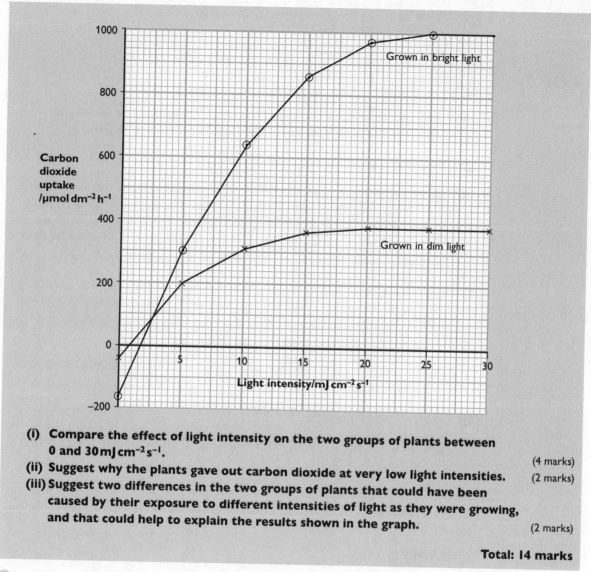

(i) **Compare the effect of light intensity on the two groups of plants between 0 and 30 mJ cm^{-2} s^{-1}.**

(4 marks)

(ii) **Suggest why the plants gave out carbon dioxide at very low light intensities.** (2 marks)

(iii) **Suggest two differences in the two groups of plants that could have been caused by their exposure to different intensities of light as they were growing, and that could help to explain the results shown in the graph.**

(2 marks)

Total: 14 marks

Part (a) is straightforward, but (b) and (c) will require a lot of careful thought. Read all the information at least twice before you try to think out good answers to the questions.

Student A

(a) (i) E ✓
(ii) D ✓

ⓔ Both correct. **2/2**

(b) It would not be able to make any ATP, ✓ which is needed for the Calvin cycle. ✓ Without the Calvin cycle, it would not be able to make carbohydrates.

ⓔ This is correct, but lacking in detail. **2/4**

(c) (i) Neither of the groups took up any carbon dioxide when there was no light. ✓ Then the quantity of carbon dioxide increased dramatically for the bright light plants, and slowly for the dim light plants. Then it levelled out, lower for the dim light plants than for the bright light ones. ✓ The dim light plants levelled out at 380 and the bright light ones at $1000\,\mu mol\,dm^{-2}\,h^{-1}$. ✓

ⓔ There are some good comparative points made here. A fundamental error, however, is that the student expresses his or her answer as though the x axis showed time — for example using the word 'slowly' and 'then'. It is also not a good idea to use terms such as 'dramatically'. See Student B for a better way of expressing these points. **3/5**

(ii) They could not photosynthesise, ✓ so the carbon dioxide in their leaves just went back out into the air again.

ⓔ One correct point is made here. **1/2**

(iii) The ones that had grown in the bright light could have bigger leaves and more chlorophyll. ✓

ⓔ The suggestion about bigger leaves is not correct. Even if the plants did have bigger leaves, this would not affect the results, because the carbon dioxide uptake is measured per unit area (look at the units on the y axis of the graph). The second point is a good suggestion. **1/2**

Student B

(a) (i) E ✓
(ii) D ✓

ⓔ Both correct. **2/2**

(b) No ATP would be made ✓ in the light-dependent reaction, so there would not be any available for the light-independent reactions, where carbohydrates (triose phosphate) are made in the Calvin cycle ✓. ATP is needed to convert GP to triose phosphate ✓ (along with reduced NADP) and also to help regenerate RuBP ✓ from the triose phosphate so the cycle can continue.

ⓔ All correct and with good detail. **4/4**

(c) (i) Below about 0.5 light intensity, both groups gave out carbon dioxide. ✓ As light intensity increased, the amount of carbon dioxide taken up by the plants grown in bright light increased more steeply ✓ than for the ones grown in dim light. In the group grown in dim light, the maximum rate of carbon dioxide uptake was $380\,\mu mol\,dm^{-2}\,h^{-1}$, whereas for the ones in bright light it was much higher, ✓ at $1000\,\mu mol\,dm^{-2}\,h^{-1}$. ✓ For the bright light plants, the maximum rate of photosynthesis was not reached until the light intensity was $25\,mJ\,cm^{-2}\,s^{-1}$, but for the ones grown in dim light the maximum rate was reached at a lower ✓ light intensity of $20\,mJ\,cm^{-2}\,s^{-1}$.

ⓔ A good answer, with some comparative figures quoted (with their units). Note the avoidance of any vocabulary that could be associated with time. **5/5**

(ii) When the light intensity was very low, the plants would not be able to photosynthesise so they would not take up any carbon dioxide. ✓ However, they would still be respiring (they respire all the time) so their leaf cells would be producing carbon dioxide ✓ which would diffuse out into the air. Normally, this carbon dioxide would be taken up by the cells for photosynthesis.

ⓔ A good answer. **2/2**

(iii) The plants grown in the light would probably be a darker green because they would have more chlorophyll ✓ in their chloroplasts, so they would be able to absorb more light and photosynthesise faster. They might also have more chloroplasts in each palisade cell. ✓ And leaves sometimes produce an extra layer of palisade cells if they are in bright light. ✓

ⓔ This answer actually contains three points, and two of them have been explained, which was not required. Student B could have got 2 marks with a much shorter answer. Nevertheless, this answer shows good understanding of the underlying biology. **2/2**

Question 4

A study was carried out in Japan to investigate how climate had changed in the past 300 000 years. As part of the study, drilling took place to a depth where the sediments were known to be about 300 000 years old, and a core was removed. The numbers and types of pollen grains present in the sediments of different ages in the core were recorded. The bottom diagram shows some of the results.

The top diagram shows probable global temperatures over this time, inferred from other data.

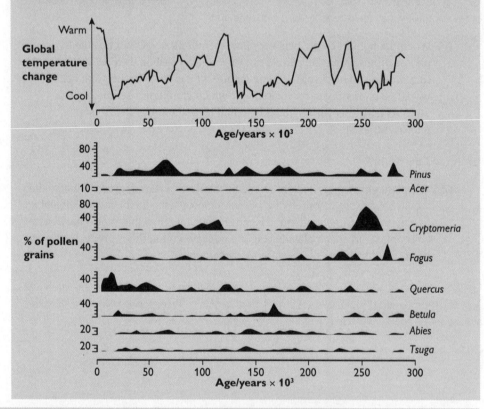

(a) *Tsuga*, *Abies*, *Pinus* and *Betula* are trees that grow best in cold climates. Discuss the extent to which the results from the pollen grain study show correlation with the changes in global temperature obtained from other data. (6 marks)

(b) Describe two possible types of the 'other data' used to draw conclusions about global temperatures in the past 300 000 years, as shown in the top diagram. (4 marks)

Total: 10 marks

e Part (a) is another question where you need to give yourself plenty of time to read and understand the information you have been given. Part (b) is easier — but note there are 2 marks for each description.

Student A

(a) There is a correlation between the cold periods and the abundance of the four named species of trees that like cold climates. For example, *Betula* has peaks at about 25, 175 and 250 × 10³ years, which matches roughly with when there were cold periods. ✓ And there are not many Betula when it was hot, ✓ for example at 115 and 200 × 10³ years ago. Similar patterns are shown by *Tsuga*, *Abies* and *Pinus*.

e This answer is good as far as it goes, but it needs to contain more information in order to get closer to full marks. Student A could have described the pattern of one of the other cold-climate trees in more detail, and could also have picked out some data that do not show such a good correlation between cold temperatures and the abundance of these tree species. **2/6**

(b) Dendrochronology is the study of tree rings, which grow at different thicknesses depending on the climate. We can match up tree rings from wood of different ages, so even if a piece of wood is very old we can probably work out its age ✓ if its tree ring pattern overlaps with a more modern piece. When the rings are wide, that means the year was warm. ✓
Carbon dioxide in air trapped in ice cores can be measured. ✓ We can assume that the more carbon dioxide there was, the warmer Earth's temperature. ✓ The deeper the ice, the older it is.

e Just enough for 2 marks on each of the methods. **4/4**

Student B

(a) The global climate seems to have had warm peaks at around 0, 125, and 220 thousand years ago. ✓ *Pinus* does seem to have not grown very much at those times. For example, there are not many *Pinus* at time 0, but there are a lot of them just before that when it was colder. ✓ But sometimes there have been lots of *Pinus* when it was warm, like at about 280 thousand years ago. ✓

e This is a difficult question, because there is a very large amount of data and it is not easy to match up the times when there are peaks on one of the graphs with another. Student B has made a reasonable attempt. He or she has identified the main warm periods, and has made some clear statements about the abundance of one of the cold climate trees at those times. The answer also attempts to 'discuss' the issue, because there are statements supporting the idea of the correlation and also one that does not support it. **3/6**

(b) Tree rings. ✓ Trees put on a new ring of growth in their trunks each year, so if you cut down a tree you can count the rings to find out how old it is. In a good year, the rings will be wider than in a bad year, so you can work out what the climate was by looking at the thickness of the rings. ✓
Ice cores. You can dig up long columns of ice from Greenland and look at the air bubbles in them to find out how much carbon dioxide there was at different depths, which means different times ago. ✓ The more carbon dioxide, the higher the temperature probably was (because of the greenhouse effect). ✓

e A good answer. Student B has learnt this part of his or her work well, though he or she does not always express answers clearly. For example, what is meant by a 'good' year? **4/4**

Unit 4: The Natural Environment and Species Survival

Question 5

The diagram shows the bacterium *Mycobacterium tuberculosis*, which causes tuberculosis (TB).

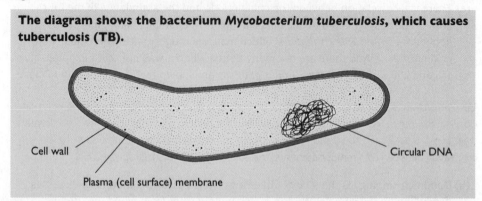

Cell wall

Circular DNA

Plasma (cell surface) membrane

(a) List three ways in which the structure of this bacterium differs from that of a virus.

(3 marks)

(b) *M. tuberculosis* is taken up by macrophages, and multiplies inside them. Explain how this strategy helps to protect *M. tuberculosis* from the immune response by B cells.

(4 marks)

(c) In an experiment to investigate how *M. tuberculosis* avoids destruction by macrophages, bacteria were added to a culture of macrophages obtained from the alveoli of mice. At the same time, a quantity of small glass beads, equivalent in size to the bacteria, were added to the culture.
The experiment was repeated using increasing quantities of bacteria and glass beads.
After 4 hours, the macrophages were sampled to find out how many had taken up either glass beads or bacteria. The results are shown in the graph. The x axis shows the initial ratio of bacteria or glass beads to macrophages in the mixture.

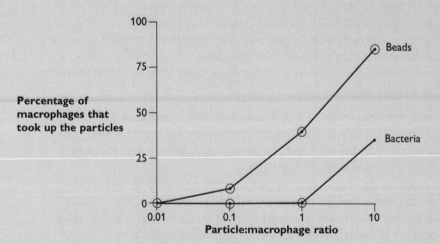

Discuss what these results suggest about the ability of macrophages to take up *M. tuberculosis*.

(3 marks)

(d) When *M. tuberculosis* is present inside a phagosome of a macrophage, it secretes glycolipids that accumulate in lysosomes and prevent them fusing with the phagosome. Explain how this prevents the macrophage from destroying the bacterium.

(3 marks)

(e) With reference to your answer to (d), suggest how natural selection could result in a population of mice that are resistant to *M. tuberculosis*.

(6 marks)

Total: 19 marks

ⓔ The most demanding parts of this question are (c) and (e) — (c) because you will have to think carefully about the data in the graph, and (e) because you need at least 6 strong points in your answer.

Student A

(a) It is made of a cell, ✓ it has a cell wall ✓ and it is much larger. ✓

ⓔ Three correct differences. **3/3**

(b) It stops the B cells seeing them, so they don't make antibodies ✓ against them.

ⓔ This is not a very clear answer. B cells do not 'see', so this is not a good term to use. The 'they' in the second sentence could refer to either B cells or the bacteria. **I/4**

(c) The macrophages took up more glass beads than bacteria. ✓ So they are not very good at taking up the bacteria. ✓

ⓔ Just enough for 2 marks, although the second sentence is weak. **2/3**

(d) Lysosomes contain digestive enzymes, ✓ so if they don't fuse with the phagosome the bacteria won't get digested. ✓

ⓔ Once again, Student A has the right ideas, but does not give enough biological detail to get full marks. **2/3**

(e) If a mouse had a gene that made a chemical that stopped the bacteria secreting the glycolipids, ✓ or if it had a gene that made the lysosomes fuse with the phagosomes even if the lipids were there, ✓ that would give it an advantage and it would be more likely to survive. ✓

ⓔ This answer explains how a mouse could have a selective advantage, but it does not go on to explain how a population of resistant mice could arise. **3/6**

Student B

(a) The bacterium is cellular, but a virus is not. ✓ A virus has a capsid made of protein, but a bacterium does not. ✓ A bacterium has a cell wall, but a virus does not. ✓

ⓔ Three correct points. **3/3**

(b) B cells only become active when they meet the specific antigen ✓ to which they are able to respond. If the bacteria are inside a macrophage, then the B cell's receptors won't meet the antigen ✓ on the bacteria. This means that the B cells will not divide to produce plasma cells, ✓ and will not secrete antibodies ✓ against the bacteria.

ⓔ A good answer. **4/4**

(c) The cells only started to take up any bacteria when the particle:macrophage ratio was 1. ✓ On the other hand, they took up glass beads even when the ratio was only just above 0.01. ✓ When the ratio of particles to macrophages was 10, only about 30% ✓ of the macrophages had taken up bacteria, whereas over 75% of them had taken up glass beads. ✓ This shows the macrophages do take up the bacteria, but not as well as they take up glass beads. ✓

ⓔ A good answer, which does attempt to 'discuss' by providing statements relating to the relatively low ability of the macrophages to take up the bacteria, but also stating that they do take them up. **3/3**

(d) Normally, lysosomes fuse with phagosomes and release hydrolytic enzymes ✓ into them. These enzymes then hydrolyse (digest) whatever is in the phagosome. ✓ If this doesn't happen, then the bacteria can live inside the phagosome ✓ without being digested.

ⓔ All correct. **3/3**

(e) Not all mice will have the same alleles, ✓ so some mice might happen to have an allele of a gene that means its lysosomes are not sensitive to the glycolipids that the bacteria produce. ✓ These mice would be able to destroy the bacteria, ✓ so they would not get TB. ✓ They would be more likely to survive and reproduce, ✓ and pass on their alleles to their offspring. ✓ This could happen over several generations, so the advantageous alleles would become more common, ✓ until perhaps the whole population had these alleles and was resistant to the bacteria.

ⓔ This answer works logically through the mechanism of natural selection, applied to this particular situation. It is clearly expressed. **6/6**

Question 6

(a) Describe how you could use quadrats and a transect to investigate the distribution and abundance of a named species in a named habitat.   (6 marks)

(b) (i) Explain the difference between *abiotic factors* and *biotic factors*, giving one example of each.   (3 marks)

(ii) It has been stated that, at the beginning of a succession, abiotic factors are more important than biotic factors. However, as succession proceeds, biotic factors become more important.
With reference to an example of succession that you have studied, discuss this statement.   (5 marks)

Total: 14 marks

ⓔ In (b) (ii) you should give specific examples from *one* particular example of succession. Note that you are asked to 'discuss' the statement, so try to give at least one point that disagrees with it, as well as some that support it.

Student A

(a) On a rocky sea shore, you could investigate limpets. ✓ Get a long tape and stretch it from down by the water up the shore. ✓ Get a quadrat and put it down next to the tape next to the water. Count the number of limpets in the quadrat ✓ and write it down. Then put the quadrat down again a bit further up the quadrat and count the limpets again. Keep doing this until you get to the end of the tape.

ⓔ Student A appears to have done an investigation like this, and he or she clearly knows how to use a transect and quadrat to measure distribution and abundance. In the fifth sentence, the student has written 'quadrat' where he or she really meant 'tape'. It is important to reread your answers if you have time. There is quite a bit of important detail missing. For example, at what intervals along the tape should the quadrats be placed? How many would you use? Would you do any repeats? **3/6**

(b) (i) Abiotic means not to do with living organisms, like salinity. ✓ Biotic means to do with living organisms, ✓ like competition. ✓

ⓔ All correct. **3/3**

(ii) On a sandy sea shore, you get succession to sand dunes. ✓ To start with, it is all just sand and it is difficult to live in because the sand doesn't hold water and there is nowhere for animals to live because there aren't many plants. ✓ Marram grass can start to grow there and its underground stems hold the sand together. There aren't any other plants so there isn't much competition. ✓ Later on soil builds up, so lots more plants can live there and they start to compete with one another. ✓

Ⓔ Again, this answer suggests that Student A has studied succession, and knows some detail about a particular example. The description is correct, but it is not well tailored to the question, and does not even mention the words 'biotic' or 'abiotic'. However, there are four good points made. **4/5**

Student B

(a) I investigated ox-eye daisies in a field margin next to a hedge. ✓ I got a 30 m tape and spread it out so that one end was at the edge of the hedge and the other end was in the crop, ✓ so the tape stretched right across the margin. I got a 1 m² quadrat ✓ and put it at distance 0 ✓ on the tape and counted the number of ox-eye daisy plants inside it. ✓ Then I moved the quadrat along by 2 metres ✓ and did the same. I kept doing this until I got to the end of the tape. Then I moved along the margin a bit and put the tape down again, and repeated the survey. ✓ I calculated the average ✓ number of ox-eye daisy plants at each distance from the hedge and drew a graph of number of plants against distance from hedge. ✓

Ⓔ This is an excellent answer, and it shows that Student B has done this investigation. There is plenty of important detail, such as a sensible size of quadrat to use and how far apart the quadrats were placed. **6/6**

(b) (i) Biotic factors are environmental factors that affect living organisms ✓ and are caused by other living organisms, such as parasites or predators. ✓ Abiotic factors are ones that are not to do with living organisms, ✓ such as light intensity. ✓

Ⓔ All correct. **3/3**

(ii) When a glacier retreats, it leaves bare ground. ✓ It is really difficult to live there because of abiotic factors, such as lack of soil and exposure to wind. ✓ However, once the pioneer plants ✓ such as mountain avens ✓ have been there for a while, they improve the soil structure ✓ and make nitrates that other plants can use. So now the abiotic factors are not so harsh and lots of different species can start to live there, ✓ so competition ✓ between them gets more intense. So I think the statement is correct.

Ⓔ A specific example is given, and there is plenty of detail about particular factors and species at different stages of the succession. **5/5**

Question 7

(a) Describe how gel electrophoresis can be used to separate DNA fragments of different lengths.

(6 marks)

(b) The diagram shows the results of electrophoresis on DNA samples taken from a mother, her child and its alleged father.

Explain what can be concluded from these results.

(3 marks)

Total: 9 marks

ⓔ This is a straightforward question, but take care to write enough in (a) to give you a good chance of getting all 6 marks.

Student A

(a) First you cut the DNA up into pieces using restriction enzymes. ✓ Then you put the DNA onto some agarose gel ✓ in a tank and switch on the power supply so the DNA gets pulled along the gel. ✓ The bigger pieces move more slowly, so they end up not so far along the gel as the smaller pieces. ✓ You can't see the DNA so you need to stain it with something so it shows up.

ⓔ There are no errors in this answer, but a little more detail is needed. **4/6**

(b) The child has one band that is in its mother's DNA and another that is in the alleged father's DNA. ✓ So he could be the father. ✓

ⓔ Two correct points made. **2/3**

Student B

(a) The DNA is cut into fragments using restriction enzymes, ✓ which cut it at particular base sequences. Then you place samples of the DNA into little wells in agarose gel ✓ in an electrophoresis tank. A voltage is then applied ✓ across the gel. The DNA pieces have a small negative charge so they steadily move towards the positive ✓ terminal. The larger they are, the more slowly they move ✓ so the smaller ones travel further ✓ than the big ones. After a time, the power is switched off so the DNA stops moving. You can tell where it is by using radioactivity, ✓ so the DNA shows up as bands on a photographic film.

🅔 All correct and enough detail for full marks. **6/6**

(b) The top band for the child matches the top band for the mother, ✓ and the bottom band for the child matches the bottom band for the father. ✓ So the alleged father could be the child's father, ✓ though we can't be 100% certain of that because there could be another man who has this band as well. ✓

🅔 A clear and thorough conclusion. **3/3**

Edexcel A2 Biology

Sample Paper 2

Question 1

The diagram shows some of the energy transfers in a food chain. The figures are in $kJ\,m^{-2}\,year^{-1}$.

(a) (i)   Name the main process responsible for the heat loss to the atmosphere from the producers. (1 mark)

(ii) Calculate the net primary productivity of the producers. Show your working. (2 marks)

(iii) Calculate the efficiency of energy transfer from the primary consumers to the secondary consumers. Show your working. (2 marks)

(iv) The efficiency of energy transfer from producers to primary consumers is less than that from primary consumers to secondary consumers. Suggest an explanation for this. (3 marks)

(b) Many of the decomposers in this food chain are micro-organisms. Describe their role in the recycling of carbon in the ecosystem. (5 marks)

Total: 13 marks

ⓔ Set your calculations out clearly, as that gives you a good chance of getting marks even if you make a mistake and arrive at an incorrect final answer.

Student A

(a) (i) Metabolism

ⓔ Incorrect. The question requires a particular metabolic reaction. **0/1**

(ii) NPP = GPP – R
    so NPP = 20000 – 15000 = 5000 ✓

ⓔ Correct working, but no units. **1/2**

(iii) $90/1000 \times 100$ ✓ = 9% ✓

ⓔ All correct. **2/2**

(iv) Not all parts of plants are eaten, ✓ so this makes it very inefficient.

ⓔ Student A has the right idea, but has not given enough information for more than one mark. **1/3**

(b) The decomposers break down dead bodies and release their nutrients into the soil. Then they respire ✓ so they send carbon dioxide back into the air. ✓

ⓔ The first sentence is correct, but it needs to refer to carbon in order to answer the question. **2/5**

Student B

(a) (i) Respiration ✓

ⓔ Correct. **1/1**

(ii) NPP = GPP − R
  = 20000 − 15000 ✓ = 5000 kJ m^{-2} year^{-1} ✓

ⓔ Correct. **2/2**

(iii) efficiency of energy transfer = $90/1000 \times 100$ ✓ = 9% ✓

ⓔ All correct. **2/2**

(iv) Primary consumers are herbivores, and they cannot eat all of the plants because parts of the plants are underground. ✓ So there is energy in the plants that never gets into the herbivores. ✓ Also, cellulose is really difficult to digest, ✓ so a lot of the energy in the plants the herbivores eat does not get into their bodies but is lost in the faeces (which the decomposers feed on). ✓ But secondary consumers are carnivores, and they can eat all the parts of the primary consumers ✓ and they don't have cellulose in them so they are easier to digest.

ⓔ Good, but Student B has written much more than a 3 mark question requires. **3/3**

(b) They break down carbon-containing molecules ✓ from dead organisms, like carbohydrates and proteins. ✓ They use some of them to release energy by respiration, ✓ so the carbon goes back into the air as carbon dioxide. ✓

ⓔ Having written an overlong answer to (a)(iv), Student B has now not written quite enough for a 5 mark question. What is here is correct, but more could have been said, for example some examples of decomposers (bacteria or fungi) and that they also break down waste materials (such as faeces). **4/5**

Question 2

(a) **The diagram shows a short length of a DNA molecule during transcription.**

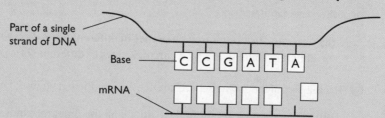

(i) **On the diagram, write single letters to represent the six bases on the mRNA strand that is being constructed.**

(2 marks)

(ii) **On the diagram, draw a ring around one codon.**

(1 mark)

(iii) **State the part of an animal cell in which transcription takes place.**

(1 mark)

(b) **Many lengths of DNA (genes) are made up of alternating exons and introns, and these are all used as a template to make an mRNA molecule with a complementary base sequence.**

(i) **Describe how this mRNA molecule is modified before it is used in the production of a protein molecule.**

(2 marks)

(ii) **Antibodies are immunoglobulins, which are protein molecules. Explain how the post-transcriptional modification of mRNA allows several different forms of an immunoglobulin to be coded for by just one gene.**

(3 marks)

(iii) **Explain why having a very large number of different immunoglobulins is advantageous to a person.**

(4 marks)

Total: 13 marks

 This question covers two different parts of the specification — protein synthesis and immunology. You should always be prepared to make links between different areas of biology.

Student A

(a) (i)

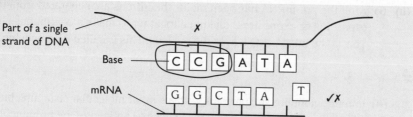

 Student A has forgotten that RNA contains U instead of T. **1/2**

(ii) See diagram

 No, codons are found on the mRNA molecule, not on DNA. **0/1**

(iii) Ribosome ✗

ⓔ No, this is incorrect. Student A has perhaps confused transcription and translation. **0/1**

(b) (i) The introns are chopped out ✓ and the exons join up with one another. ✓

ⓔ Just enough for both marks. **2/2**

(ii) You can stick the introns together in different ways ✓ so you can get lots of different mRNAs ✓ which can make lots of different sorts of antibodies.

ⓔ The last part of the answer repeats the question, so there is no mark for that. **2/3**

(iii) You can make an antibody to attack all the different kinds of pathogens that get into the body. ✓

ⓔ Correct, but nowhere near enough information for a 4 mark question. **1/4**

Student B

(a) (i)

Part of a single strand of DNA

Base — C C G A T A

✓

mRNA — G G C U A U ✓✓

ⓔ Correct. **2/2**

(ii) See diagram

ⓔ Correct. **1/1**

(iii) Nucleus ✓

ⓔ Correct. **1/1**

(b) (i) After the pre-mRNA has been made, the introns are separated from the exons. ✓ The exons are then linked together to make a continuous chain. ✓ This is the mRNA that leaves the nucleus. This is called RNA splicing.

ⓔ Correct. **2/2**

(ii) Immunoglobulins are proteins with a similar molecular structure, but with a variable region at one end which sticks to antigens. One immunoglobulin gene can make many different versions of the protein because the exons can be linked together in different ways, ✓ so the base sequence on the mRNA is different ✓ and therefore so is the amino acid sequence ✓ in the protein that is made.

ⓔ A clear and correct answer. **3/3**

(iii) Each immunoglobulin is able to bind with one particular antigen. ✓ As each antigen has a different molecular structure, you need many kinds of immunoglobulin to be sure you have one that can bind with whatever antigen ✓ gets into the body. When the immunoglobulin binds with the antigen, it helps other cells like macrophages to destroy it. ✓ So this protects you against infections ✓ and allows you to survive.

ⓔ All correct. **4/4**

Question 3

Clostridium difficile is a bacterium that lives naturally in the alimentary canal of many people. However, in people who are weakened by illness, or who have been taking antibiotics which have affected their normal gut bacterial populations, C. difficile populations in the gut may increase. The bacterium produces toxins that may cause serious diarrhoea and inflammation of the colon. In severe cases, death may result.

(a) The graph shows the number of reported cases of infection with C. difficile in hospitals in England between 1990 and 2005.
   (i) Describe the changes in the number of cases of reported infection with C. difficile between 1990 and 2005.

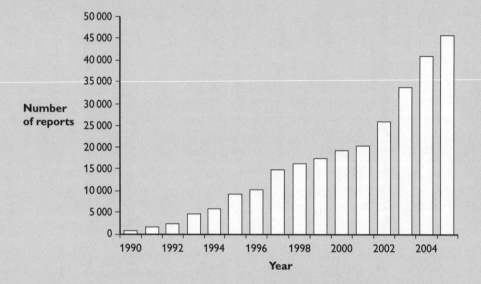

(2 marks)
   (ii) Suggest two reasons for the changes you have described. (2 marks)

(b) Mild cases of C. difficile infection are usually treated with the antibiotic metronidazole, which inhibits bacterial DNA synthesis. More severe cases are treated with the antibiotic vancomycin, which inhibits the formation of cross-links in the bacterial cell wall. Both of these antibiotics are bacteriocidal if given in sufficiently large doses.
   (i) Suggest why neither metronidazole nor vancomycin have harmful effects on human cells. (2 marks)
   (ii) Explain the meaning of the term bacteriocidal. (1 mark)
   (iii)Vancomycin is usually given intravenously (into a blood vessel), but for C. difficile infections it is normally given by mouth. Suggest why this is done. (1 mark)

Edexcel A2 Biology

(c) Since the data shown in the graph above were collected, there has been some reduction in the number of cases of *C. difficile* contracted by patients in hospitals. These are thought to be due to implementation of a code of practice for preventing and treating these infections.

Suggest how each of these guidelines in the code of practice may have helped to reduce the number of *C. difficile* infections in hospitals.

  (i)   All healthcare workers should wash their hands with soap and water before and after contact with patients with suspected *C. difficile* infection.   (2 marks)

  (ii)   Antibiotics used for the treatment of *C. difficile* infection should not be broad-spectrum ones (that is, antibiotics that kill a wide range of bacteria), but should be specifically targeted at *C. difficile*, and antibiotics should not be used at all unless clearly necessary.   (3 marks)

Total: 13 marks

ⓔ There is a lot of reading in this question. Read *all* of it, very carefully, before you begin to construct your answers.

Student A

(a) (i)   There has been a fairly steady rate of increase ✓ since 1990, from about 250 to just over 45 000 a year. This is 180 times ✓ as many cases.

ⓔ Two correct statements, including some manipulation of the data. 'Fairly steady' is just worth a mark, although a careful consideration of the shape of the graph actually shows that the rate of increase is greater in more recent years. **2/2**

   (ii)   People are living longer, so there are more old people around and they are the ones most likely to get infected. ✓ And people don't wash their hands properly in hospitals.

ⓔ The first suggestion is valid, but the second one (although it may well be an important factor in spreading the bacterium) cannot explain the change in incidence, unless people wash their hands less now than they used to. **1/2**

(b) (i)   Because our cells are different from bacteria.

ⓔ There is not enough detail here for any marks to be awarded. **0/2**

   (ii)   Kills bacteria. ✓

ⓔ Correct. **1/1**

   (iii) Because the bacterium infects the alimentary canal, so it is quicker for the anitbiotic to reach it. ✓

ⓔ Correct. **1/1**

(c) (i)   This would remove any bacteria from their hands ✓ so they won't give them to another patient when they touch them. ✓

ⓔ Just enough for both marks. **2/2**

> **(ii)** It's more likely that the antibiotics will kill *C. difficile*.

ⓔ Student A is thinking on the right lines, but the answer needs to contain more information. **0/3**

Student B

> **(a) (i)** The incidence has risen by about 42 000 cases over a 15 year period, a mean increase of 2800 cases a year. ✓ The rise has been steeper since 2001. ✓

ⓔ Two good points, one involving some manipulation of the figures. **2/2**

> **(ii)** There may be more bacteria that have become resistant to antibiotics, ✓ so there are more *C. difficile* around that have not been killed and this could increase the rate of infection. And other treatments in hospitals are better now than they used to be, so there are more ill people in hospitals who might have died before, and they have weak immune systems so they can't fight off the infection. ✓

ⓔ Two good suggestions, well explained. **2/2**

> **(b) (i)** The enzymes involved in the synthesis of DNA in bacteria are probably different ✓ from humans. And human cells don't have cell walls. ✓

ⓔ Two correct points, specifically related to the information about these two antibiotics given in the question. **2/2**

> **(ii)** It is something that kills bacteria. ✓

ⓔ Correct. **1/1**

> **(iii)** The bacteria are in the alimentary canal, so the antibiotic will get to them quicker. ✓

ⓔ Correct. **1/1**

> **(c) (i)** This removes bacteria from their hands, ✓ especially if they use soap. So if they have picked up bacteria from one person, they will not transmit them ✓ to another one.

ⓔ A good suggestion, well explained. **2/2**

> **(ii)** *C. difficile* is more likely to grow if there aren't other 'friendly' bacteria in the digestive system, so if you use broad-spectrum antibiotics you might kill the friendly bacteria ✓ and actually make the infection worse. ✓ By not using antibiotics at all you reduce the risk of resistant strains evolving. ✓

ⓔ Two good points, one of them well explained. **3/3**

Question 4

The polymerase chain reaction, PCR, is widely used in many branches of gene technology.

(a) Describe the purpose of using the polymerase chain reaction.   (2 marks)

(b) During the polymerase chain reaction, the temperature of the reacting mixture is repeatedly changed, allowing different enzymes to work.
(i) Outline the main stages of the polymerase chain reaction.   (6 marks)
(ii) Explain why enzymes work best at a particular temperature.   (5 marks)

Total: 13 marks

ⓔ Part (b) (i) asks you to 'outline', so you are not expected to write about each stage in detail — just enough to make clear that you know the sequence of steps and can summarise each one.

Student A

(a) To amplify the DNA. ✓

ⓔ Correct, but not enough for 2 marks. **1/2**

(b) (i) The temperature starts off very hot, then it is cooled and then it is heated up again.

ⓔ Not enough here for any marks. **0/6**

(ii) They work faster when it gets warmer, then when it goes above their optimum temperature they get denatured ✓ and stop working.

ⓔ This is a description (and not a very good one) not an explanation — except for the mention of denaturation. **1/5**

Student B

(a) If you have only a tiny amount ✓ of DNA, you can make millions of copies ✓ of it so you have enough to do DNA profiling. ✓

ⓔ Brief but absolutely correct. **2/2**

(b) (i) First the DNA is split apart into two strands ✓ by heating it to 95 °C. ✓ Then primers are attached ✓ at a temperature of 65 °C. ✓ Then DNA polymerase ✓ starts at the primers and makes a copy of each strand of the DNA, ✓ at a temperature of 72 °C. ✓ Then the new DNA molecules are heated up and split apart again, and so on.

ⓔ A short answer that is full of correct detail. **6/6**

(ii) Enzymes don't work very fast at low temperatures because they don't have much kinetic energy ✓ so they don't bump into their substrate very often. ✓ But when it is too hot the hydrogen bonds ✓ in the enzyme molecule break so the molecule loses its shape ✓ and the substrate doesn't fit in the active site. ✓ In between there is the perfect temperature for reacting.

🅮 Another short answer that is packed with relevant facts. **5/5**

Question 5

(a) Carbon dioxide is a greenhouse gas.
    (i) Name one other greenhouse gas.   (1 mark)
    (ii) Explain what is meant by the term greenhouse gas.   (3 marks)

(b) The graph shows the quantity of carbon dioxide emitted by vehicles in an area of Australia between 1990 and 2008, and the predicted emissions up until 2020.

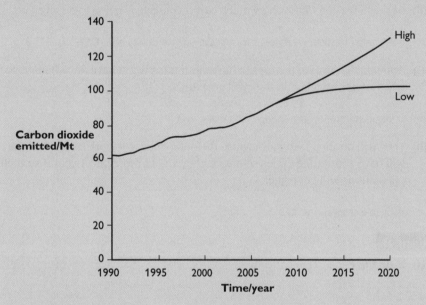

    (i) Suggest reasons for the increase in emissions of carbon dioxide shown in the graph.   (2 marks)
    (ii) Explain why it is very difficult to predict the shape of the curve beyond the present year.   (2 marks)

(c) Discuss the possible effects of global warming on each of the following.
    (i) The distribution and abundance of organisms that are adapted to living in cold climates, such as on high mountains.   (3 marks)
    (ii) The development and life cycles of insects.   (3 marks)

Total: 14 marks

ⓔ Part (c) is a 'discuss' question, so you should try to include more than one idea in each of your answers.

Student A

(a) (i) Methane ✓

ⓔ Correct. **1/1**

(ii) It is a gas that traps heat ✓ and is making the Earth hotter.

ⓔ A very 'thin' answer. **1/3**

(b) (i) Maybe more people are driving cars ✓ or they are driving them further. ✓

ⓔ Two sensible suggestions, although dangerously close to being too vague to get both marks. **2/2**

(ii) Because we don't know how much carbon dioxide is going to be given off.

ⓔ This is not a helpful answer — it needs to suggest why we don't know. **0/2**

(c) (i) Animals that live on the tops of high mountains won't have anywhere to go when it gets hotter, ✓ so they will go extinct.

ⓔ A rather vague statement; just enough for a mark. **1/3**

(ii) They will develop faster ✓ because their enzymes will work faster. And they will have shorter life cycles so maybe they will fit more into 1 year ✓ so their populations will get bigger.

ⓔ Two reasonable suggestions. **2/3**

Student B

(a) (i) Methane ✓

ⓔ Correct. **1/1**

(ii) It is a gas in the atmosphere ✓ that lets short wavelength rays from the Sun get through and down to the surface, but traps the long wavelength rays ✓ from escaping. So they heat up the air ✓ or bounce back down to the surface and make the Earth warmer.

ⓔ All correct. **3/3**

(b) (i) The population has increased, ✓ so there are more cars on the road. ✓ People are more affluent, so they make more car journeys or take more flights in aeroplanes. ✓

ⓔ Good suggestions. **2/2**

(ii) It depends on what governments do about trying to reduce carbon dioxide emissions. ✓ The 'high' curve is probably if we just go on as we are, and the 'low' one if enough countries do something about emitting less carbon dioxide. And we don't know if the oceans or tropical rainforests might be able to absorb more ✓ of the carbon dioxide, or how many rainforests are going to be cut down. ✓

ⓔ A succinct summary of some of the main difficulties. **2/2**

> **(c)** **(i)** It is cold on top of a mountain and things that live there are adapted ✓ to live in cold places. If it gets warmer they may not be able to survive. ✓ And species that need warmer climates might be able to live ✓ on the tops of mountains so the first species would have to compete ✓ with them and they might not be good at that. So we would expect the abundance of the mountain organisms to decrease.

ⓔ Some good ideas, well expressed. **3/3**

> **(ii)** Aphids might have more than one life cycle a year, ✓ so their population might increase. ✓ They will develop faster at higher temperatures. ✓ Some of them might be able to live in places that were too cold for them before. ✓

ⓔ Again, several good suggestions. **3/3**

Question 6

(a) The diagram shows the Calvin cycle.

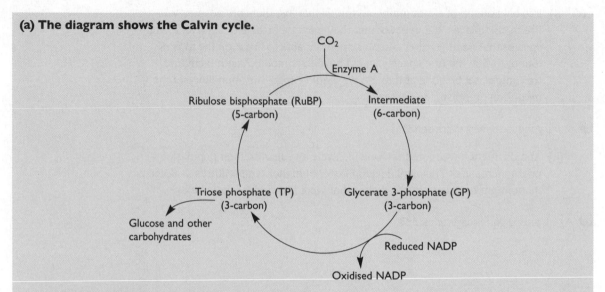

(i) **Name enzyme A.** (1 mark)

(ii) **On the diagram, draw and label arrows to show how and where ATP from the light-dependent reaction is used.** (2 marks)

(iii) **From the diagram, what is the first carbohydrate made during the Calvin cycle?** (1 mark)

(b) An experiment was carried out in which photosynthesising tissues were exposed to light, then to darkness and then to light again. The graph shows the relative amounts of GP (glycerate 3-phosphate) and RuBP (ribulose bisphosphate) during the experiment.

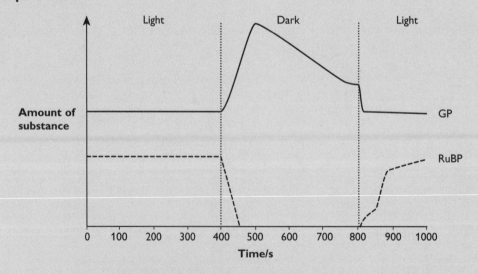

(i) **Describe the effects of light and dark on the amounts of GP and RuBP.** (4 marks)

(ii) **Suggest explanations for the changes shown in the graph.** (5 marks)

Total: 13 marks

ⓔ The hardest part of this question is (b) (ii). Look for a change in gradient of one of the lines, and think of a logical cause of this change. Then do the same for another change in gradient, until you feel you have made at least five strong points.

Student A

(a) (i) Rubisco ✓

ⓔ Correct. **1/1**

(ii)

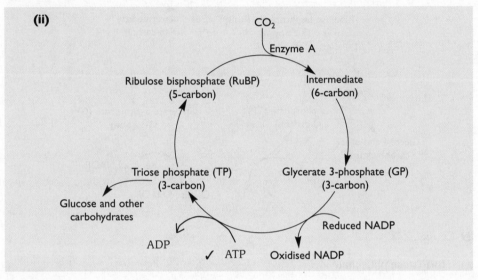

ⓔ One correct, but there should be another one. **1/2**

(iii) Ribulose bisphosphate ✗

ⓔ Incorrect. **0/1**

(b) (i) When it is light, they stay the same. In the dark, GP increases but RuBP decreases. ✓ When it is light again they go back to normal.

ⓔ A poorly worded answer, lacking detail. 'Stay the same' could mean the same as each other, or constant. 'Go back to normal' does not indicate what 'normal' might be. **1/4**

(ii) When it is light, the plant can do the light-dependent reactions ✓ and the light-independent reactions, so it can make plenty of RuBP and GP in the Calvin cycle. When it is dark it runs out of reduced NADP and ATP ✓ made in the light-dependent reactions, so it can't change GP to triose phosphate ✓ so the GP just builds up. ✓

ⓔ This is a much better answer than the previous one. Student A works logically through the probable reasons for the patterns shown on the graphs. However, he or she has not addressed the fall in RuBP during darkness, or the gradual fall in GP after it has risen during darkness. **4/5**

Student B

(a) (i) RUBISCO ✓

ⓔ Correct. 1/1

(ii)

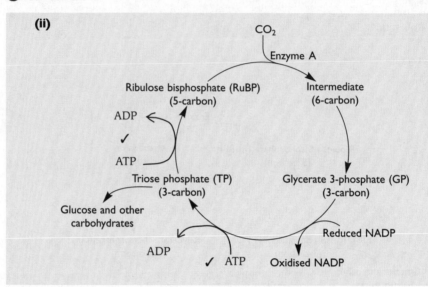

ⓔ Correct. 2/2

(iii) Triose phosphate ✓

ⓔ Correct. 1/1

(b) (i) During the light, the amounts of GP and RuBP remain constant. ✓ When it is dark, the GP quickly increases to a peak ✓ and then steadily decreases. ✓ The RuBP quickly disappears altogether. ✓ When the light comes back on, the GP drops quickly ✓ and then goes back to the level it was at the start of the experiment and stays there. ✓ The RuBP immediately starts to increase ✓ until it reaches its original level. ✓

ⓔ A thorough description. 4/4

(ii) In the dark, there won't be any ATP or reduced NADP ✓ from the light-dependent reaction. ✓ So the Calvin cycle gets stuck at GP ✓ because it can't be turned into triose phosphate, ✓ so the GP just builds up. As there is no ATP, the triose phosphate (what there is of it, because no more is being made now) can't be changed into RuBP ✓ so this decreases to nothing ✓ as it is all made into GP. When the RuBP all runs out, no more GP is made so it stops increasing. ✓ Maybe some of the GP just disintegrates if it can't be made into triose phosphate which is why it falls.

ⓔ Well thought out and clearly explained. 5/5

Question 7

A group of islands contains three species of mice, each species being found on only one island. A fourth species is found on the mainland. A region of the DNA of each species was sequenced, and the percentage differences between the samples were calculated. The results are shown in the table.

	Mainland	Island A	Island B
Island A	6.1	–	–
Island B	4.8	9.7	–
Island C	5.2	10.3	7.5

(a) Discuss how these results suggest that the species of mouse on each island has evolved from the species on the mainland, and not from one of the other island species. *(5 marks)*

(b) Each of these four species of mice is unable to breed with any of the other species, even if they are placed together.
Suggest how reproductive isolation between the mice could have arisen, and explain its role in speciation. *(6 marks)*

Total: 11 marks

ⓔ Note that (b) asks you to do two things — firstly 'suggest', and then 'explain'.

Student A

(a) The three island mice each have DNA more similar to the mainland mouse than to each other. ✓✓ So they have probably all evolved from the mainland mouse. If one of the island mice had evolved from another island mouse, their DNA would be more similar. ✓

ⓔ This answer shows that Student A has managed to work out what the table shows, but it does not provide enough detail in the discussion to get all of the marks available. **3/5**

(b) If they are on different islands, they will have different selection pressures ✓ so the mice might end up different. They might have different courtship behaviour, ✓ so they won't be able to mate with each other. ✓ You have to get reproductive isolation to produce a new species.

ⓔ A reasonable description. The last sentence is moving towards another mark, but it really only repeats what is already in the question. **3/6**

Student B

(a) From the table, we can see that each island mouse's DNA is more similar to the DNA of the mainland mouse than to any of the other island mice. ✓ For example, the island C mice have DNA that is 10.3% different from the island A mice and 7.5% different from the island B mice, but only 5.2% different from the mainland mice. ✓ The longer ago two species split away from each other, the more different

we would expect their DNA to be. ✓ This is because the longer the time, the more mutations ✓ are likely to have occurred, so there will be different base sequences ✓ in the DNA.

e A good answer to a difficult question. Student B has made good use of the data in the table, and has used some of the figures to support his or her answer. The last part of the answer explains why differences in DNA base sequence indicate degree of relationship. **5/5**

(b) A species is defined as a group of organisms that can interbreed with each other to produce fertile offspring. ✓ So to get new species you have to have something that stops them reproducing together so genes can't flow ✓ from one species to the other. This might happen if different selection pressures ✓ acted on two populations of a species, so that different alleles were selected for ✓ and over many generations their genomes became more different. ✓ So they might be the wrong size and shape ✓ to be able to breed with each other.

e The answer begins with a clearly explained link between speciation and reproductive isolation, and then goes on to describe how two populations could become reproductively isolated. **6/6**

Knowledge check answers

1 Both ATP and an RNA nucleotide are made up of ribose, a base and phosphate. ATP has three phosphate groups, whereas a nucleotide has only one. An RNA nucleotide may have any one of four bases, whereas in ATP the base is always adenine.

2 The oxygen released in photosynthesis comes from water molecules, not carbon dioxide.

3 The energy is transferred to ATP and reduced NADP.

4 The substrates are carbon dioxide and RuBP, and the product is GP.

5 They are produced in the light-dependent reactions.

6 It is transferred to decomposers when the feed on the detritus.

7 energy in primary consumers (herbivorous insects) = 300
energy in secondary consumers (spiders) = 30
so efficiency = (30 ÷ 300) × 100 = 10%

8 A habitat is the type of place in which an organism lives. A community is all the living organisms, of all species, that live together in the same place at the same time.

9 Photosynthesis is the only process that takes inorganic carbon dioxide from the atmosphere and produces organic (carbon-containing) compounds that can be used by other organisms.

10 Using the mean temperature over a 30-year period gives an informative base-line against which any changes can be clearly seen.

11 The changes relate to the seasons. In summer, when sunlight is brighter and days are longer, there is more photosynthesis, so more carbon dioxide is removed from the atmosphere. In winter there is less photosynthesis, so carbon dioxide levels in the atmosphere rise.

12 The youngest ring on the tree (i.e. the last one to be made when the tree was growing) is the one on the right hand side of the sample, which dates to somewhere later than the year 1200 — perhaps 1300. (The diagram does not allow us to be any more precise than this, but in practice the wood could probably be dated to within a year or so.)

13 The graph is showing differences in surface temperature from the mean temperature in the year 2000.

14 We only have their structures to compare — and often even these can be only partially reconstructed. We know nothing about their physiology, behaviour or breeding habits. We cannot know if they could reproduce together to produce fertile offspring.

15 There is only room in the DNA double helix for one nucleotide with one base, and one with two bases, to link together. Moreover, C and G join with three hydrogen bonds, while A and T join with two.

16 Three

17 GUA, CCU, GAC

18 The primer provides the DNA polymerase with a starting point for building the new DNA strand.

19 nucleus, mitochondria, RER, SER, Golgi apparatus, centrioles, lysosomes

20 Transcription is the synthesis of a molecule of mRNA using DNA as a template. Reverse transcriptase catalyses the synthesis of a strand of DNA using RNA as a template.

21 An antibody is a protein (an immunoglobulin) secreted by B cells in response to a specific antigen. An antibiotic is a substance that is made outside the body (for example by a fungus, or synthesised in a laboratory) that is taken as a medicine to kill bacteria inside the body.

22 By diffusion

Index